Cliffs Quick Review

Algebra I

by
Jerry Bobrow, Ph.D.

Consultants

Pam Mason, M.A.
Ron Podrasky, M.A.

FIRST EDITION

ISBN 0-8220-5302-0

CONTENTS

PRE-ALGEBRA

PRELIMINARIES AND BASIC OPERATIONS 3
Preliminaries 3
Groups of numbers 3
Ways to show multiplication 4
Common math symbols 5
Properties of Basic Mathematical Operations 5
Some properties (axioms) of addition 5
Some properties (axioms) of multiplication 6
A property of two operations 8
Multiplying and Dividing Using Zero 8
Powers and Exponents 9
Squares and cubes 10
Operations with powers and exponents 10
Square Roots and Cube Roots 12
Square roots 12
Cube roots 13
Approximating square roots 13
Simplifying square roots 14
Grouping Symbols: Parentheses, Brackets, Braces 15
Parentheses ( ) 15
Brackets [ ] and braces { } 15
Order of operations 16
Divisibility Rules 17

SIGNED NUMBERS AND FRACTIONS 19
Signed Numbers (Positive Numbers and Negative Numbers) 19
Number lines 19
Addition of signed numbers 19
Subtraction of signed numbers 20
Minus preceding parenthesis 21
Multiplying and dividing signed numbers 21

Fractions 22
Negative fractions 22
Adding positive and negative fractions 22
Subtracting positive and negative fractions 23
Multiplying fractions 24
Canceling 24
Multiplying mixed numbers 25
Dividing fractions 25
Dividing complex fractions 26
Dividing mixed numbers 27
Simplifying Fractions and Complex Fractions 27
Decimals 28
Changing terminating decimals to fractions 28
Changing fractions to decimals 29
Changing infinite repeating decimals to fractions 30
Important equivalents that can save you time 32
Percent 33
Changing decimals to percents 33
Changing percents to decimals 33
Changing fractions to percents 34
Changing percents to fractions 34
Finding percent of a number 34
Other applications of percent 35
Percent—proportion method 36
Scientific Notation 37
Multiplication in scientific notation 38
Division in scientific notation 39

ALGEBRA

TERMINOLOGY, SETS, AND EXPRESSIONS 43
Some Basic Language.......................... 43
Understood multiplication.................... 43
Letters to be aware of 43
Set Theory 43
Special sets 43
Describing sets.............................. 44
Types of sets................................ 44
Comparing sets............................... 44
Operations with sets 44
Variables and Algebraic Expressions 45
Evaluating Expressions....................... 46

EQUATIONS, RATIOS, AND PROPORTIONS 49
Equations.................................... 49
Axioms of equality 49
Solving equations 49
Literal equations 53
Ratios and Proportions....................... 55
Ratios 55
Proportions 56
Solving proportions for value 57

EQUATIONS WITH TWO VARIABLES........... 59
Solving Systems of Equations (Simultaneous Equations) .. 59
Addition/subtraction method.................. 59
Substitution method 62
Graphing method 63

MONOMIALS, POLYNOMIALS, AND FACTORING 65
Monomials 65
Adding and subtracting monomials 65
Multiplying monomials 66
Dividing monomials 67
Working with negative exponents 68
Polynomials 69
Adding and subtracting polynomials 69
Multiplying polynomials 70
Dividing polynomials by monomials 72
Dividing polynomials by polynomials 73
Factoring 76
Factoring out a common factor 76
Factoring the difference between two squares 76
Factoring polynomials having three terms of the form $ax^2 + bx + c$ 77
Factoring by grouping 82

ALGEBRAIC FRACTIONS 83
Operations with Algebraic Fractions 83
Reducing algebraic fractions 83
Multiplying algebraic fractions 84
Dividing algebraic fractions 85
Adding or subtracting algebraic fractions 86

INEQUALITIES, GRAPHING, AND ABSOLUTE VALUE 89
Inequalities 89
Axioms and properties of inequalities 89
Solving inequalities 90

Graphing on a Number Line 92
Graphing inequalities 92
Intervals 94
Absolute Value 95
Solving equations containing absolute value 95
Solving inequalities containing absolute value and graphing 97

ANALYTIC GEOMETRY **101**
Coordinate Graphs 101
Graphing equations on the coordinate plane 103
Slope and intercept of linear equations 106
Graphing linear equations using slope and intercept .. 109
Finding the equation of a line 112
Linear Inequalities and Half-Planes 115
Open half-plane 115
Closed half-plane 117

FUNCTIONS AND VARIATIONS **119**
Functions 119
Relations 119
Domain and range 119
Defining a function 120
Graphs of functions 121
Graphs of relationships that are not functions 122
Determining domain, range, and if the relation is a function 123
Finding the values of functions 124
Variations 126
Direct variation 126
Inverse variation (indirect variation) 128

ROOTS AND RADICALS 131
Simplifying Square Roots 131
Operations with Square Roots 133
Under a single radical sign 133
When radical values are alike 133
When radical values are different 134
Addition and subtraction of square roots after simplifying 134
Products of nonnegative roots 135
Quotients of nonnegative roots 136

QUADRATIC EQUATIONS 139
Solving Quadratic Equations 139
Factoring 139
The quadratic formula 142
Completing the square 146

WORD PROBLEMS 151
Solving Technique 151
Key Words and Phrases 152
Simple Interest 154
Compound Interest 154
Ratio and Proportion 155
Percent 156
Percent Change 157
Number Problems 158
Age Problems 161
Motion Problems 163
Coin Problems 165
Mixture Problems 167
Work Problems 169
Number Problems with Two Variables 170

PRE-ALGEBRA

Reviewing some basics and getting up to speed. Algebra is often defined as the branch of mathematics in which letters are used to represent numbers and signs are used to represent operations: a kind of universal arithmetic or simply arithmetic using letters. Before you can really understand algebra, you need to feel comfortable with basic mathematical operations and what is commonly called pre-algebra. The following pre-algebra sections review some of the essentials to get you ready for an easy transition into the world of algebra.

Preliminaries

Groups of numbers. In doing algebra, you will work with several groups of numbers.

- **Natural or counting numbers.** The numbers 1, 2, 3, 4, . . . are called **natural** or **counting numbers.**

- **Whole numbers.** The numbers 0, 1, 2, 3, . . . are called **whole numbers.**

- **Integers.** The numbers . . . −2, −1, 0, 1, 2, . . . are called **integers.**

- **Negative integers.** The numbers . . . −3, −2, −1 are called **negative integers.**

- **Positive integers.** The natural numbers are sometimes called the **positive integers.**

- **Rational numbers.** Fractions, such as $\frac{3}{2}$ or $\frac{7}{8}$, are called **rational numbers.** Since a number such as 5 may be written as 5/1, all *integers* are *rational numbers.* All rational numbers can be written as fractions a/b, with a being an integer and b being a natural number. Terminating and repeating decimals are also rational numbers, since they can be written as fractions in this form.

- **Irrational numbers.** Another type of number is an **irrational number.** Irrational numbers *cannot* be written as fractions a/b, with a being an integer and b being a natural number. $\sqrt{3}$ and π are examples of irrational numbers.

- **Prime numbers.** A **prime number** is a number that has exactly two factors, or that can be evenly divided by only itself and 1. For example, 19 is a prime number because it can be evenly divided by only 19 and 1, but 21 is not a prime number because 21 can be evenly divided by other numbers (3 and 7). The only even prime number is 2; thereafter, any even number may be divided evenly by 2. Zero and 1 are *not* prime numbers. The first ten prime numbers are 2, 3, 5, 7, 11, 13, 17, 19, 23, and 29.

- **Odd numbers. Odd numbers** are whole numbers not divisible by 2: 1, 3, 5, 7, . . .

- **Even numbers. Even numbers** are numbers divisible by 2: 0, 2, 4, 6, . . .

- **Composite numbers.** A **composite number** is a number divisible by more than just 1 and itself: 4, 6, 8, 9, . . .

- **Squares. Squares** are the result when numbers are multiplied by themselves: $(2 \cdot 2 = 4)$, $(3 \cdot 3 = 9)$; 1, 4, 9, 16, 25, 36, . . .

- **Cubes. Cubes** are the result when numbers are multiplied by themselves twice: $(2 \cdot 2 \cdot 2 = 8)$, $(3 \cdot 3 \cdot 3 = 27)$: 1, 8, 27, . . .

Ways to show multiplication. There are several ways to show multiplication. They are

$$4 \times 3 = 12$$

$$4 \cdot 3 = 12$$

$$(4)(3) = 12$$

$$4(3) = 12$$

$$(4)3 = 12$$

Common math symbols. Symbol references:

$=$ is equal to
$\neq$ is not equal to
$>$ is greater than
$<$ is less than
$\geqq$ is greater than or equal to (also written $\geq$)
$\leqq$ is less than or equal to (also written $\leq$)
$\ngtr$ is not greater than
$\nless$ is not less than
$\ngeqq$ is not greater than or equal to
$\nleqq$ is not less than or equal to

Properties of Basic Mathematical Operations

Some properties (axioms) of addition.

- **Closure** is when all answers fall into the original set. If you add two even numbers, the answer is still an even number ($2 + 4 = 6$); therefore, the set of even numbers *is closed* under addition (has closure). If you add two odd numbers, the answer is not an odd number ($3 + 5 = 8$); therefore, the set of odd numbers is *not closed* under addition (no closure).

- **Commutative** means that the *order* does not make any difference.

$$2 + 3 = 3 + 2$$

$$a + b = b + a$$

Note: Commutative does *not* hold for subtraction.

$$3 - 1 \neq 1 - 3$$

$$a - b \neq b - a$$

- **Associative** means that the *grouping* does not make any difference.

$$(2 + 3) + 4 = 2 + (3 + 4)$$

$$(a + b) + c = a + (b + c)$$

The grouping has changed (parentheses moved), but the sides are still equal.

Note: Associative does *not* hold for subtraction.

$$4 - (3 - 1) \neq (4 - 3) - 1$$

$$a - (b - c) \neq (a - b) - c$$

- The **identity element** for addition is 0. Any number added to 0 gives the original number.

$$3 + 0 = 3$$

$$a + 0 = a$$

- The **additive inverse** is the opposite (negative) of the number. Any number plus its additive inverse equals 0 (the identity).

$3 + (-3) = 0$; therefore, 3 and -3 are additive inverses

$-2 + 2 = 0$; therefore, -2 and 2 are additive inverses

$a + (-a) = 0$; therefore, a and $-a$ are additive inverses

Some properties (axioms) of multiplication.

- **Closure** is when all answers fall into the original set. If you multiply two even numbers, the answer is still an even number $(2 \times 4 = 8)$; therefore, the set of even numbers *is closed* under multiplication (has closure). If you multiply two odd numbers, the answer is an odd number $(3 \times 5 = 15)$;

therefore, the set of odd numbers *is closed* under multiplication (has closure).

- **Commutative** means that the *order* does not make any difference.

$$2 \times 3 = 3 \times 2$$

$$a \times b = b \times a$$

Note: Commutative does *not* hold for division.

$$2 \div 4 \neq 4 \div 2$$

- **Associative** means that the *grouping* does not make any difference.

$$(2 \times 3) \times 4 = 2 \times (3 \times 4)$$

$$(a \times b) \times c = a \times (b \times c)$$

The grouping has changed (parentheses moved) but the sides are still equal.

Note: Associative does *not* hold for division.

$$(8 \div 4) \div 2 \neq 8 \div (4 \div 2)$$

- The **identity element** for multiplication is 1. Any number multiplied by 1 gives the original number.

$$3 \times 1 = 3$$

$$a \times 1 = a$$

- The **multiplicative inverse** is the **reciprocal** of the number. Any number multiplied by its reciprocal equals 1.

$2 \times \frac{1}{2} = 1$; therefore, 2 and $\frac{1}{2}$ are multiplicative inverses

$a \times 1/a = 1$; therefore, a and $1/a$ are multiplicative inverses (provided $a \neq 0$)

A property of two operations. The **distributive property** is the process of distributing the number on the outside of the parentheses to each term on the inside.

$$2(3 + 4) = 2(3) + 2(4)$$

$$a(b + c) = a(b) + a(c)$$

Note: You cannot use the distributive property with only one operation.

$$3(4 \times 5 \times 6) \neq 3(4) \times 3(5) \times 3(6)$$

$$a(bcd) \neq a(b) \times a(c) \times a(d) \text{ or } (ab)(ac)(ad)$$

Multiplying and Dividing Using Zero

Zero times any number equals zero.

$$0 \times 5 = 0$$

$$0 \times (-3) = 0$$

$$8 \times 9 \times 3 \times (-4) \times 0 = 0$$

Likewise, zero divided by any number is zero.

$$0 \div 5 = 0$$

$$\frac{0}{3} = 0$$

$$0 \div (-6) = 0$$

Important note: Dividing by zero is "undefined" and is not permitted.

$$\frac{6}{0} \text{ is not permitted}$$

because there is no such answer. The answer *is not* zero.

Powers and Exponents

An **exponent** is a positive or negative number placed above and to the right of a quantity. It expresses the **power** to which the quantity is to be raised or lowered. In 4^3, 3 is the exponent. It shows that 4 is to be used as a factor three times. $4 \times 4 \times 4$ (multiplied by itself twice). 4^3 is read as *four to the third power* (or *four cubed* as discussed below).

$$2^4 = 2 \times 2 \times 2 \times 2 = 16$$

$$3^2 = 3 \times 3 = 9$$

Remember that $x^1 = x$ and $x^0 = 1$ when x is any number (other than 0).

$$2^1 = 2 \qquad 2^0 = 1$$

$$3^1 = 3 \qquad 3^0 = 1$$

$$4^1 = 4 \qquad 4^0 = 1$$

If the exponent is negative, such as 3^{-2}, then the number and exponent may be dropped under the number 1 in a fraction to remove the negative sign.

Example 1: Simplify the following by removing the exponents.

(a) $3^{-2} = \dfrac{1}{3^2} = \dfrac{1}{9}$

(b) $2^{-3} = \dfrac{1}{2^3} = \dfrac{1}{8}$

(c) $3^{-4} = \dfrac{1}{3^4} = \dfrac{1}{81}$

Squares and cubes. Two specific types of powers should be noted, **squares** and **cubes.** To *square a number,* just multiply it by itself (the exponent would be 2). For example, 6 squared (written 6^2) is 6×6, or 36. 36 is called a **perfect square** (the square of a whole number). Following is a list of perfect squares.

$0^2 = 0$ $5^2 = 25$ $9^2 = 81$

$1^2 = 1$ $6^2 = 36$ $10^2 = 100$

$2^2 = 4$ $7^2 = 49$ $11^2 = 121$

$3^2 = 9$ $8^2 = 64$ $12^2 = 144$ etc.

$4^2 = 16$

To *cube a number,* just multiply it by itself twice (the exponent would be 3). For example, 5 cubed (written 5^3) is $5 \times 5 \times 5$, or 125. 125 is called a **perfect cube** (the cube of a whole number). Following is a list of perfect cubes.

$0^3 = 0$ $4^3 = 64$

$1^3 = 1$ $5^3 = 125$

$2^3 = 8$ $6^3 = 216$

$3^3 = 27$ $7^3 = 343$ etc.

Operations with powers and exponents. To *multiply* two numbers with exponents, *if the base numbers are the same,* simply keep the base number and add the exponents.

Example 2: Multiply the following, leaving the answers with exponents.

(a) $2^3 \times 2^5 = 2^8 \quad (2 \times 2 \times 2) \times (2 \times 2 \times 2 \times 2 \times 2) = 2^8$

(b) $3^2 \times 3^4 = 3^6$

To *divide* two numbers with exponents, *if the base numbers are the same,* simply keep the base number and subtract the second exponent from the first, or the exponent of the denominator from the exponent of the numerator.

Example 3: Divide the following, leaving the answers with exponents.

(a) $4^8 \div 4^5 = 4^3$

(b) $\frac{9^6}{9^2} = 9^4$

To *multiply* or *divide* numbers with exponents, *if the base numbers are different,* you must simplify each number with an exponent first and then perform the operation.

Example 4: Simplify and perform the operation indicated.

(a) $3^2 \times 2^2 = 9 \times 4 = 36$

(b) $6^2 \div 2^3 = 36 \div 8 = 4\frac{4}{8} = 4\frac{1}{2}$

(Some shortcuts are possible.)

To *add* or *subtract* numbers with exponents, *whether the base numbers are the same or different,* you must simplify each number with an exponent first and then perform the indicated operation.

Example 5: Simplify and perform the operation indicated.

(a) $3^2 - 2^3 = 9 - 8 = 1$

(b) $4^3 + 3^2 = 64 + 9 = 73$

If a *number with an exponent is taken to another power* $(4^2)^3$, simply keep the original base number and multiply the exponents.

Example 6: Multiply and leave the answers with exponents.

(a) $(4^2)^3 = 4^6$

(b) $(3^3)^2 = 3^6$

Square Roots and Cube Roots

Note that **square** and **cube roots** and operations with them are often included in algebra sections, and the following will be discussed further in that section.

Square roots. To find the **square root** of a number, you want to find some number that when multiplied by itself gives you the original number. In other words, to find the square root of 25, you want to find the number that when multiplied by itself gives you 25. The square root of 25, then, is 5. The symbol for square root is $\sqrt{\ }$. Following is a list of perfect (whole number) square roots.

$$\sqrt{0} = 0 \qquad \sqrt{16} = 4 \qquad \sqrt{64} = 8$$
$$\sqrt{1} = 1 \qquad \sqrt{25} = 5 \qquad \sqrt{81} = 9$$
$$\sqrt{4} = 2 \qquad \sqrt{36} = 6 \qquad \sqrt{100} = 10 \quad \text{etc.}$$
$$\sqrt{9} = 3 \qquad \sqrt{49} = 7$$

Special note: If no sign (or a positive sign) is placed in front of the square root, then the positive answer is required. Only if a negative sign is in front of the square root is the negative answer required. This notation is used in many texts and will be adhered to in this book. Therefore,

$$\sqrt{9} = 3 \qquad \text{and} \qquad -\sqrt{9} = -3$$

Cube roots. To find the **cube root** of a number, you want to find some number that when multiplied by itself twice gives you the original number. In other words, to find the cube root of 8, you want to find the number that when multiplied by itself twice gives you 8. The cube root of 8, then, is 2, since $2 \times 2 \times 2 = 8$. Notice that the symbol for cube root is the radical sign with a small three (called the **index**) above and to the left $\sqrt[3]{\ }$. Other roots are similarly defined and identified by the index given. (In square root, an index of two is understood and usually not written.) Following is a list of **perfect** (whole number) **cube roots.**

$$\sqrt[3]{0} = 0 \qquad \sqrt[3]{27} = 3$$
$$\sqrt[3]{1} = 1 \qquad \sqrt[3]{64} = 4$$
$$\sqrt[3]{8} = 2 \qquad \sqrt[3]{125} = 5$$

Approximating square roots. To find the square root of a number that is not a perfect square, it will be necessary to find an *approximate* answer by using the procedure given in Example 7.

Example 7: Approximate $\sqrt{42}$.

The $\sqrt{42}$ is between $\sqrt{36}$ and $\sqrt{49}$.

$$\sqrt{36} < \sqrt{42} < \sqrt{49}$$

$$\sqrt{36} = 6$$

$$\sqrt{49} = 7$$

Therefore, $6 < \sqrt{42} < 7$, and since 42 is almost halfway between 36 and 49, $\sqrt{42}$ is almost halfway between $\sqrt{36}$ and $\sqrt{49}$. To check, multiply: $6.5 \times 6.5 = 42.25$, or about 42.

Square roots of nonperfect squares can be approximated, looked up in tables, or found by using a calculator. You may wish to keep these two in mind:

$$\sqrt{2} \cong 1.414 \qquad \sqrt{3} \cong 1.732$$

Simplifying square roots. Sometimes you will have to *simplify* square roots, or write them in simplest form. In fractions, $\frac{2}{4}$ can be reduced to $\frac{1}{2}$. In square roots, $\sqrt{32}$ can be simplified to $4\sqrt{2}$. To *simplify a square root,* first factor the number under the $\sqrt{\ }$ into two factors, one of which is the largest possible perfect square. (Perfect square numbers are 1, 4, 9, 16, 25, 49, . . .)

Example 8: Simplify.

$$\sqrt{32} = \sqrt{16 \times 2}$$

Then take the square root of the perfect square number.

$$\sqrt{16 \times 2} = \sqrt{16} \times \sqrt{2} = 4 \times \sqrt{2}$$

and finally write as a single expression.

$$4\sqrt{2}$$

Remember that most square roots cannot be simplified, as they are already in simplest form, such as $\sqrt{7}$, $\sqrt{10}$, $\sqrt{15}$.

Grouping Symbols: Parentheses, Brackets, Braces

Parentheses ( ). Parentheses are used to group numbers or variables. Everything inside parentheses must be done before any other operations.

Example 9: Simplify.

$$50(2 + 6) = 50(8) = 400$$

When a parenthesis is preceded by a minus sign, to remove the parentheses, change the sign of each term within the parentheses.

Example 10: Simplify.

$$6 - (-3 + a - 2b + c) =$$
$$6 + 3 - a + 2b - c =$$
$$9 - a + 2b - c$$

Brackets [ ] and braces { }. Brackets and **braces** are also used to group numbers or variables. Technically, they are used after parentheses. Parentheses are to be used first, then brackets, then braces: {[( )]}. Sometimes, instead of brackets or braces, you will see the use of larger parentheses.

$$\Big((3 + 4) \cdot 5\Big) + 2$$

A number using all three grouping symbols would look like this.

$$2\{1 + [4(2 + 1) + 3]\}$$

Example 11: Simplify. Notice that you work from the inside out.

$$\begin{aligned} 2\{1 + [4(2 + 1) + 3]\} &= \\ 2\{1 + [4(3) + 3]\} &= \\ 2\{1 + [12 + 3]\} &= \\ 2\{1 + [15]\} &= \\ 2\{16\} &= 32 \end{aligned}$$

Order of operations. If multiplication, division, powers, addition, parentheses, etc., are all contained in one problem, the **order of operations** is as follows.

1. parentheses
2. powers and square roots
3. multiplication } whichever comes first left to right
4. division }
5. addition } whichever comes first left to right
6. subtraction }

Example 12: Simplify the following problems.

(a) $6 + 4 \times 3 =$

$6 + 12 =$ (multiplication)

18 (then addition)

(b) $10 - 3 \times 6 + 10^2 + (6 + 1) \times 4 =$

$10 - 3 \times 6 + 10^2 + (7) \times 4 =$ (parentheses first)

$10 - 3 \times 6 + 100 + (7) \times 4 =$ (powers next)

$10 - 18 + 100 + 28 =$ (multiplication)

$-8 + 100 + 28 =$ (addition/subtraction left to right)

$92 + 28 = 120$

An easy way to remember the order of operations *after parentheses* is: **P**lease **M**y **D**ear **A**unt **S**arah (**P**owers, **M**ultiplication, **D**ivision, **A**ddition, **S**ubtraction).

Divisibility Rules

The following set of rules can help you save time in trying to check the divisibility of numbers.

A number is divisible by	if
2	it ends in 0, 2, 4, 6, or 8
3	the sum of its digits is divisible by 3
4	the number formed by the last two digits is divisible by 4
5	it ends in 0 or 5
6	it is divisible by 2 and 3 (use the rules for both)
7	(no simple rule)
8	the number formed by the last three digits is divisible by 8
9	the sum of its digits is divisible by 9

Example 13:

(a) Is 126 divisible by 3? Sum of digits = 9. Since 9 is divisible by 3, then 126 is divisible by 3.

(b) Is 1,648 divisible by 4? Since 48 is divisible by 4, then 1,648 is divisible by 4.

(c) Is 186 divisible by 6? Since 186 ends in 6, it is divisible by 2. Sum of digits = 15. Since 15 is divisible by 3, 186 is divisible by 3. 186 is divisible by 2 and 3; therefore, it is divisible by 6.

(d) Is 2,488 divisible by 8? Since 488 is divisible by 8, then 2,488 is divisible by 8.

(e) Is 2,853 divisible by 9? Sum of digits = 18. Since 18 is divisible by 9, then 2,853 is divisible by 9.

Signed Numbers (Positive Numbers and Negative Numbers)

Number lines. On a **number line,** numbers to the right of 0 are positive. Numbers to the left of 0 are negative, as shown in Figure 1.

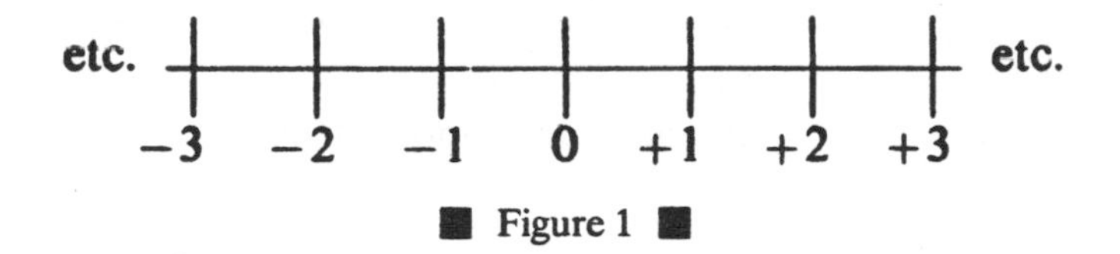

■ Figure 1 ■

Given any two numbers on a number line, the one on the right is always larger, regardless of its sign (positive or negative). Note that fractions may also be placed on a number line as shown in Figure 2.

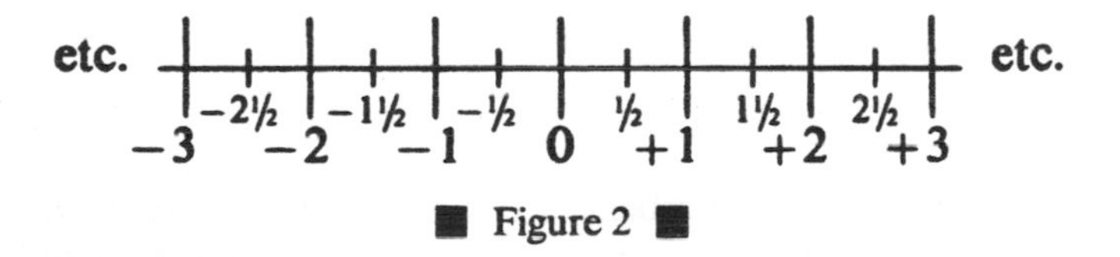

■ Figure 2 ■

Addition of signed numbers. When *adding two numbers with the same sign* (either both positive or both negative), add the numbers and keep the same sign.

Example 1: Add the following.

$$\text{(a)}\quad \begin{array}{r} +5 \\ \underline{+\ +7} \\ +12 \end{array} \qquad \text{(b)}\quad \begin{array}{r} -8 \\ \underline{+\ -3} \\ -11 \end{array}$$

When *adding two numbers with different signs* (one positive and one negative), subtract the numbers and keep the sign from the larger one.

Example 2: Add the following.

$$\text{(a)}\quad \begin{array}{r} +5 \\ \underline{+\ -7} \\ -2 \end{array} \qquad \text{(b)}\quad \begin{array}{r} -59 \\ \underline{+\ +72} \\ +13 \end{array}$$

Signed numbers may also be added "horizontally."

Example 3: Adding the following.

(a) $+9 + 6 = +15$

(b) $-12 + 9 = -3$

(c) $8 + (-5) = 3$

Subtraction of signed numbers. To *subtract positive and/or negative numbers,* just change the sign of the number being subtracted and then add.

Example 4: Subtract the following.

$$\text{(a)}\quad \begin{array}{r} +12 \\ \underline{-\ +4} \\ \\ +12 \\ \underline{+\ -4} \\ +8 \end{array} \qquad \text{(b)}\quad \begin{array}{r} -14 \\ \underline{-\ -4} \\ \\ -14 \\ \underline{+\ +4} \\ -10 \end{array} \qquad \text{(c)}\quad \begin{array}{r} -19 \\ \underline{-\ +6} \\ \\ -19 \\ \underline{+\ -6} \\ -25 \end{array} \qquad \text{(d)}\quad \begin{array}{r} +20 \\ \underline{-\ -3} \\ \\ +20 \\ \underline{+\ +3} \\ +23 \end{array}$$

Subtracting positive and/or negative numbers may also be done "horizontally."

Example 5: Subtract the following.

(a) $+12 - (+4) = +12 + (-4) = 8$

(b) $+16 - (-6) = +16 + (+6) = 22$

(c) $-20 - (+3) = -20 + (-3) = -23$

(d) $-5 - (-2) = -5 + (+2) = -3$

Minus preceding parenthesis. If a *minus precedes a parenthesis,* it means everything within the parentheses is to be subtracted. Therefore, using the same rule as in subtraction of signed numbers, simply change every sign within the parentheses to its opposite and then add.

Example 6: Subtract the following.

(a) $9 - (+3 - 5 + 7 - 6) =$

$9 + (-3 + 5 - 7 + 6) =$

$9 + (+1) = 10$

(b) $20 - (+35 - 50 + 100) =$

$20 + (-35 + 50 - 100) =$

$20 + (-85) = -65$

Multiplying and dividing signed numbers. To *multiply or divide signed numbers,* treat them just like regular numbers but remember this rule: An odd number of negative signs will produce a negative answer. An even number of negative signs will produce a positive answer.

Example 7: Multiply or divide the following.

(a) $(-3)(+8)(-5)(-1)(-2) = +240$

(b) $(-3)(+8)(-1)(-2) = -48$

(c) $\frac{-64}{-2} = +32$

(d) $\frac{-64}{+2} = -32$

Fractions

Negative fractions. Fractions may be *negative* as well as positive. (See the number line in Figure 2, page 19.) However, negative fractions are typically written

$-\frac{3}{4}$ not $\frac{-3}{4}$ or $\frac{3}{-4}$ (although they are all equal)

$$-\frac{3}{4} = \frac{-3}{4} = \frac{3}{-4}$$

Adding positive and negative fractions. The rules for signed numbers (page 19) apply to fractions as well.

Example 8: Add the following.

(a) $-\frac{1}{2}+\frac{1}{3}=-\frac{3}{6}+\frac{2}{6}=-\frac{1}{6}$

(b)
$$\begin{array}{rcr} +\frac{3}{4} & = & +\frac{9}{12} \\ +-\frac{1}{3} & = & +-\frac{4}{12} \\ \hline & & +\frac{5}{12} \end{array}$$

Subtracting positive and negative fractions. The rule for subtracting signed numbers (page 20) applies to fractions as well.

Example 9: Subtract the following.

(a)
$$\begin{array}{rcrcr} +\frac{9}{10} & = & +\frac{9}{10} & = & +\frac{9}{10} \\ --\frac{1}{5} & = & ++\frac{1}{5} & = & +\frac{2}{10} \\ \hline & & & & +\frac{11}{10}=1\frac{1}{10} \end{array}$$

(b) $+\frac{2}{3}-\left(-\frac{1}{5}\right)=\frac{10}{15}-\left(-\frac{3}{15}\right)=\frac{10}{15}+\frac{3}{15}=\frac{13}{15}$

(c) $+\frac{1}{3}-\frac{3}{4}=+\frac{4}{12}-\frac{9}{12}=+\frac{4}{12}+\left(-\frac{9}{12}\right)=-\frac{5}{12}$

Multiplying fractions. To *multiply fractions,* simply multiply the numerators and then multiply the denominators. Reduce to lowest terms if necessary.

Example 10: Multiply.

$$\frac{2}{3} \times \frac{5}{12} = \frac{10}{36} \qquad \text{reduce } \frac{10}{36} \text{ to } \frac{5}{18}$$

This answer had to be reduced because it wasn't in lowest terms. Since whole numbers can also be written as fractions ($3 = \frac{3}{1}$, $4 = \frac{4}{1}$, etc.), the problem $3 \times \frac{3}{8}$ would be worked by changing 3 to $\frac{3}{1}$.

Canceling. *Canceling when multiplying fractions* would have eliminated the need to reduce your answer. To cancel, find a number that divides evenly into one numerator and one denominator. In this case, 2 will divide evenly into 2 in the numerator (it goes in one time) and 12 in the denominator (it goes in six times). Thus,

$$\frac{\overset{1}{\cancel{2}}}{3} \times \frac{5}{\underset{6}{\cancel{12}}} = \frac{5}{18}$$

Remember, you may cancel only when *multiplying* fractions. The rules for multiplying signed numbers (page 21) hold here too.

Example 11: Cancel where possible and multiply.

$$\text{(a)}\quad \frac{1}{4}\times\frac{2}{7}=\frac{1}{\cancel{4}_{2}}\times\frac{\cancel{2}^{1}}{7}=\frac{1}{14}$$

$$\text{(b)}\quad \left(-\frac{3}{8}\right)\times\left(-\frac{4}{9}\right)=\left(-\frac{\cancel{3}^{1}}{\cancel{8}_{2}}\right)\times\left(-\frac{\cancel{4}^{1}}{\cancel{9}_{3}}\right)=+\frac{1}{6}$$

Multiplying mixed numbers. To *multiply mixed numbers,* first change any mixed number to an improper fraction. Then multiply as previously shown (page 24).

Example 12: Multiply.

$$3\tfrac{1}{3}\times 2\tfrac{1}{4}=\tfrac{10}{3}\times\tfrac{9}{4}=\tfrac{90}{12}=7\tfrac{6}{12}=7\tfrac{1}{2}$$

or

$$\frac{\cancel{10}^{5}}{\cancel{3}_{1}}\times\frac{\cancel{9}^{3}}{\cancel{4}_{2}}=\frac{15}{2}=7\tfrac{1}{2}$$

Change the answer, if in improper fraction form, back to a mixed number and reduce if necessary. Remember, the rules for multiplication of signed numbers (page 21) apply here as well.

Dividing fractions. To *divide fractions,* invert (turn upside down) the second fraction (the one "divided by") and multiply. Then reduce if possible.

Example 13: Divide.

(a) $\frac{1}{6} \div \frac{1}{5} = \frac{1}{6} \times \frac{5}{1} = \frac{5}{6}$

(b) $\frac{1}{9} \div \frac{1}{3} = \frac{1}{\cancel{9}_3} \times \frac{\cancel{3}^1}{1} = \frac{1}{3}$

Here too the rules for division of signed numbers (page 21) apply.

Dividing complex fractions. Sometimes a division of fractions problem may appear in the form below (these are called **complex fractions**).

Example 14: Simplify.

$$\frac{\frac{3}{4}}{\frac{7}{8}}$$

Consider the line separating the two fractions to mean "divided by." Therefore, this problem may be rewritten as

$$\frac{3}{4} \div \frac{7}{8} =$$

Now, follow the same procedure as shown in Example 13.

$$\frac{3}{4} \div \frac{7}{8} = \frac{3}{\cancel{4}_1} \times \frac{\cancel{8}^2}{7} = \frac{6}{7}$$

Dividing mixed numbers. To *divide* **mixed numbers,** first change them to improper fractions (page 25). Then follow the rule for dividing fractions (page 25).

Example 15: Divide.

$$3\tfrac{3}{5} \div 2\tfrac{2}{3} = \frac{18}{5} \div \frac{8}{3} = \frac{\overset{9}{\cancel{18}}}{5} \times \frac{3}{\underset{4}{\cancel{8}}} = \frac{27}{20} = 1\tfrac{7}{20}$$

Notice that after you invert and have a multiplication of fractions problem, you may then cancel tops with bottoms when appropriate.

Simplifying Fractions and Complex Fractions

If either numerator or denominator consists of several numbers, these numbers must be combined into one number. Then reduce if possible.

Example 16: Simplify.

(a) $$\frac{28 + 14}{26 + 17} = \frac{42}{43}$$

(b) $$\frac{\frac{1}{4} + \frac{1}{2}}{\frac{1}{3} + \frac{1}{4}} = \frac{\frac{1}{4} + \frac{2}{4}}{\frac{4}{12} + \frac{3}{12}} = \frac{\frac{3}{4}}{\frac{7}{12}} = \frac{3}{4} \div \frac{7}{12} = \frac{3}{\underset{1}{\cancel{4}}} \times \frac{\overset{3}{\cancel{12}}}{7} = \frac{9}{7} = 1\tfrac{2}{7}$$

(c) $$\frac{3 - \frac{3}{4}}{-4 + \frac{1}{2}} = \frac{2\frac{1}{4}}{-3\frac{1}{2}} = \frac{\frac{9}{4}}{-\frac{7}{2}} = \tfrac{9}{4} \div -\tfrac{7}{2} = \frac{9}{\underset{2}{\cancel{4}}} \times -\frac{\overset{1}{\cancel{2}}}{7} = -\tfrac{9}{14}$$

(d) $$\frac{1}{1+\frac{1}{1+\frac{1}{4}}} = \frac{1}{1+\frac{1}{\frac{5}{4}}} = \frac{1}{1+(1 \div \frac{5}{4})} = \frac{1}{1+(1 \times \frac{4}{5})} =$$

$$\frac{1}{1+\frac{4}{5}} = \frac{1}{1\frac{4}{5}} = \frac{1}{\frac{9}{5}} = 1 \div \frac{9}{5} = 1 \times \frac{5}{9} = \frac{5}{9}$$

Decimals

Fractions may also be written in **decimal** form (decimal fractions) as either **terminating** (for example, .3) or **infinite repeating** (for example, .666 . . .) decimals.

Changing terminating decimals to fractions. To *change terminating decimals to fractions,* simply remember that all numbers to the right of the decimal point are fractions with denominators of only 10, 100, 1000, 10,000, etc. Next, use the technique of *read it, write it,* and *reduce it.*

Example 17: Change the following to fractions in lowest terms.

(a) .8

Read it: .8 (eight tenths)

Write it: $\frac{8}{10}$

Reduce it: $\frac{4}{5}$

(b) −.07

Read it: −.07 (negative seven hundredths)

Write it: $-\frac{7}{100}$ (can't reduce this one)

All rules for signed numbers also apply to operations with decimals.

Changing fractions to decimals. To *change a fraction to a decimal,* simply do what the operation says. In other words, $\frac{13}{20}$ means 13 divided by 20. So do just that (insert decimal points and zeros accordingly).

Example 18: Change to decimals.

(a) $-\frac{13}{20}$

$$20\overline{)-13.00} = -.65 \quad \text{(quotient } -.65\text{)}$$

(b) $\frac{2}{9}$

$$9\overline{)2.000} = .222\ldots \quad \text{(quotient } .222\ldots\text{)}$$

Changing infinite repeating decimals to fractions. Infinite repeating decimals are usually represented by putting a line over (sometimes under) the shortest block of repeating decimals. This line is called a **vinculum.** So you would write

$.\overline{3}$ to indicate .333 . . .

$.\overline{51}$ to indicate .515151 . . .

$-2.1\overline{47}$ to indicate −2.1474747 . . .

Notice that only the digits under the vinculum are repeated.

Every infinite repeating decimal can be expressed as a fraction.

Example 19: Find the fraction represented by the repeating decimal $.\overline{7}$.

Let n stand for	$.\overline{7}$	or	.77777 . . .
So $10n$ stands for	$7.\overline{7}$	or	7.77777 . . .

Since $10n$ and n have the same fractional part, their difference is an integer.

$$\begin{array}{r} 10n = 7.\overline{7} \\ -\quad n = \ .\overline{7} \\ \hline 9n = 7 \end{array}$$

You can solve this problem as follows.

$$9n = 7$$

$$n = 7/9$$

So $\qquad .\overline{7} = 7/9$

Example 20: Find the fraction represented by the repeating decimal $.\overline{36}$.

Let n stand for	$.\overline{36}$	or	$.363636\ldots$
So $10n$ stands for	$3.6\overline{36}$	or	$3.63636\ldots$
and $100n$ stands for	$36.\overline{36}$	or	$36.3636\ldots$

Since $100n$ and n have the same fractional part, their difference is an integer.

$$\begin{array}{rl} 100n = & 36.\overline{36} \\ -\quad n = & .\overline{36} \\ \hline 99n = & 36 \end{array}$$

You can solve this equation as follows.

$$99n = 36$$

$$n = 36/99$$

Now, reduce $36/99$ to $4/11$.

So $$.\overline{36} = 4/11$$

Example 21: Find the fraction represented by the repeating decimal $.5\overline{4}$.

Let n stand for	$.5\overline{4}$	or	$.544444\ldots$
So $10\,n$ stands for	$5.\overline{4}$	or	$5.44444\ldots$
and $100\,n$ stands for	$54.\overline{4}$	or	$54.4444\ldots$

Since $100n$ and $10n$ have the same fractional part, their difference is an integer.

$$\begin{array}{r} 100n = 54.\overline{4} \\ -\ 10n = 5.\overline{4} \\ \hline 90n = 49 \end{array}$$

You can solve this equation as follows.

$$90n = 49$$

$$n = 49/90$$

So $$.5\overline{4} = 49/90$$

Important equivalents that can save you time. Memorizing the following can eliminate computations.

$1/100 = .01 = 1\%$
$1/10 = .1 = 10\%$
$1/5 = 2/10 = .2 = .20 = 20\%$
$3/10 = .3 = .30 = 30\%$
$2/5 = 4/10 = .4 = .40 = 40\%$
$1/2 = 5/10 = .5 = .50 = 50\%$
$3/5 = 6/10 = .6 = .60 = 60\%$
$7/10 = .7 = .70 = 70\%$
$4/5 = 8/10 = .8 = .80 = 80\%$
$9/10 = .9 = .90 = 90\%$
$1/4 = 25/100 = .25 = 25\%$
$3/4 = 75/100 = .75 = 75\%$

$1/3 = .33\frac{1}{3} = 33\frac{1}{3}\%$
$2/3 = .66\frac{2}{3} = 66\frac{2}{3}\%$
$1/8 = .125 = .12\frac{1}{2} = 12\frac{1}{2}\%$
$3/8 = .375 = .37\frac{1}{2} = 37\frac{1}{2}\%$
$5/8 = .625 = .62\frac{1}{2} = 62\frac{1}{2}\%$
$7/8 = .875 = .87\frac{1}{2} = 87\frac{1}{2}\%$
$1/6 = .16\frac{2}{3} = 16\frac{2}{3}\%$
$5/6 = .83\frac{1}{3} = 83\frac{1}{3}\%$
$1 = 1.00 = 100\%$
$2 = 2.00 = 200\%$
$3\frac{1}{2} = 3.5 = 3.50 = 350\%$

Percent

A fraction whose denominator is 100 is called a **percent**. The word *percent* means hundredths (per hundred).

So $37\% = {}^{37}/_{100}$

Changing decimals to percents. To *change decimals to percents,*

1. Move the decimal point two places to the right.
2. Insert a percent sign.

Example 22: Change to percents,

(a) $.75 = 75\%$

(b) $.05 = 5\%$

(c) $1.85 = 185\%$

Changing percents to decimals. To *change percents to decimals,*

1. Eliminate the percent sign.
2. Move the decimal point two places to the left (sometimes, adding zeros will be necessary).

Example 23: Change to decimals,

(a) $23\% = .23$

(b) $5\% = .05$

Changing fractions to percents. To *change fractions to percents,*

1. Change to a decimal.
2. Change the decimal to a percent.

Example 24: Change to percents.

(a) $\frac{2}{5} = .4 = 40\%$

(b) $\frac{5}{2} = 2.5 = 250\%$

(c) $\frac{1}{20} = .05 = 5\%$

Changing percents to fractions. To *change percents to fractions,*

1. Drop the percent sign.
2. Write over one hundred.
3. Reduce if necessary.

Example 25: Change to fractions.

(a) $60\% = \frac{60}{100} = \frac{3}{5}$

(b) $230\% = \frac{230}{100} = \frac{23}{10} = 2\frac{3}{10}$

Finding percent of a number. To *determine percent of a number,* change the percent to a fraction or decimal (whichever is easier for you) and multiply. Remember, the word *of* means multiply.

Example 26: Find the percents of these numbers.

(a) 20% of 80 =

$^{20}/_{100} \times 80 = {}^{1600}/_{100} = 16$ or $.20 \times 80 = 16.00 = 16$

(b) ½% of 18 =

$\frac{1/2}{100} \times 18 = {}^{1}/_{200} \times 18 = {}^{18}/_{200} = {}^{9}/_{100}$ or $.005 \times 18 = .09$

Other applications of percent. Turn the question word-for-word into an equation. For *what,* substitute the letter *x*; for *is*, substitute an *equal sign*; for *of*, substitute a *multiplication sign*. Change percents to decimals or fractions, whichever you find easier. Then solve the equation.

Example 27: Turn each of the following into an equation and solve.

(a) 18 is what percent of 90?

$$18 = x(90)$$
$$^{18}/_{90} = x$$
$$^{1}/_{5} = x$$
$$20\% = x$$

(b) 10 is 50% of what number?

$$10 = .50(x)$$
$$^{10}/_{.50} = x$$
$$20 = x$$

(c) What is 15% of 60?

$$x = {}^{15}\!/_{100} \times 60 = {}^{900}\!/_{100} = 9$$

or $$.15(60) = 9$$

Percent—proportion method. Another simple method commonly used to solve percent problems is the **proportion** or **is/of method.** First set up a blank proportion and then fill in the empty spaces by using the following steps.

$$\frac{?}{?} = \frac{?}{?}$$

Example 28: 30 is what percent of 50?

1. Whatever is next to the percent (%) is put over 100. (The word *what* is the unknown, or x.)

$$\frac{x}{100} = \frac{?}{?}$$

2. Whatever comes immediately after the word *of* goes on the bottom of one side of the proportion.

$$\frac{x}{100} = \frac{?}{50}$$

3. Whatever is left (comes next to the word *is*) goes on top, on one side of the proportion.

$$\frac{x}{100} = \frac{30}{50}$$

4. Then solve the proportion.

$$\frac{x}{100} = \frac{30}{50}$$

In this particular instance, it can be observed that $30/50 = 60/100$, so the answer is 60%. Solving mechanically on this problem would not be time effective.

This method works for the three basic types of percent questions.

1. 30 is what percent of 50?
2. 30 is 20% of what number?
3. What number is 30% of 50? (In this type it is probably easier to simply multiply the numbers.)

Scientific Notation

Very large or very small numbers are sometimes written in **scientific notation.** A number written in scientific notation is a number between 1 and 10 multiplied by a power of 10.

Example 29: Express the following in scientific notation.

(a) 2,100,000 written in scientific notation is 2.1×10^6. Simply place the decimal point to get a number between 1 and 10 and then count the digits to the right of the decimal to get the power of 10.

2.100000. moved 6 digits to the left

(b) .0000004 written in scientific notation is 4×10^{-7}. Simply place the decimal point to get a number between 1 and 10 and then count the digits from the original decimal point to the new one.

.0000004. moved 7 digits to the right

Notice that whole numbers have positive exponents and fractions have negative exponents.

Multiplication in scientific notation. To *multiply* numbers in *scientific notation,* simply multiply the numbers that are between 1 and 10 together to get the first number and add the powers of ten to get the second number.

Example 30: Multiply and express the answers in scientific notation.

(a) $(2 \times 10^2)(3 \times 10^4) =$

$$(2 \times 10^2)(3 \times 10^4) = 6 \times 10^6$$

(b) $(6 \times 10^5)(5 \times 10^7) =$

$$(6 \times 10^5)(5 \times 10^7) = 30 \times 10^{12}$$

This answer must be changed to scientific notation (first number from 1 to 9).

$30 \times 10^{12} = 3.0 \times 10^1 \times 10^{12} = 3.0 \times 10^{13}$

(c) $(4 \times 10^{-4})(2 \times 10^5) =$

$$(4 \times 10^{-4})(2 \times 10^5) = 8 \times 10^1$$

Division in scientific notation. To *divide* numbers in *scientific notation,* simply divide the numbers that are between 1 and 10 to get the first number and subtract the powers of ten to get the second number.

Example 31: Divide and express the answers in scientific notation.

(a) $(8 \times 10^5) \div (2 \times 10^2) =$

$$(8 \times 10^5) \div (2 \times 10^2) = 4 \times 10^3$$

(b) $\dfrac{7 \times 10^9}{4 \times 10^3} = (7 \div 4)(10^9 \div 10^3) = 1.75 \times 10^6$

(c) $(6 \times 10^7) \div (3 \times 10^9) =$

$$(6 \times 10^7) \div (3 \times 10^9) = 2 \times 10^{-2}$$

(d) $(2 \times 10^4) \div (5 \times 10^2) =$

$$(2 \times 10^4) \div (5 \times 10^2) = .4 \times 10^2$$

This answer must be changed to scientific notation.

$.4 \times 10^2 = 4 \times 10^{-1} \times 10^2 = 4 \times 10^1$

(e) $(8.4 \times 10^5) \div (2.1 \times 10^{-4}) =$

$$(8.4 \times 10^5) \div (2.1 \times 10^{-4}) = 4 \times 10^{5-(-4)} = 4 \times 10^9$$

ALGEBRA

Algebra, as mentioned earlier, is essentially arithmetic with some of the numbers replaced by letters or variables. The letters or variables are merely substitutes for numbers. Initially, algebra referred to equation solving, but now it encompasses the language of algebra and the patterns of reasoning. The rules for algebra are basically the same as the rules for arithmetic.

Some Basic Language

Understood multiplication. When two or more letters or a number and letter(s) are written next to each other, they are *understood to be multiplied.* Thus, $8x$ means 8 times x ($x8$ is never written). Or ab means a times b. Or $18ab$ means 18 times a times b.

Parentheses also represent multiplication. Thus, 3(4) means 3 times 4. A raised dot also means multiplication. Thus, $6 \cdot 5$ means 6 times 5.

Letters to be aware of. Although they may appear in some texts, we recommend that you never use o, e, or i as variables. (Technically, e and i stand for constants or predetermined numbers, and o is too easily confused with 0—zero.) When using z, you may wish to write it as ƶ so it is not confused with 2.

Set Theory

A set is a group of objects, numbers, etc.—{1, 2, 3}. An **element** is a member of a set. $3 \in \{1, 2, 3\}$. 3 is an element of the set of 1, 2, 3.

Special sets. A **subset** is a set within a set—$\{2, 3\} \subset \{1, 2, 3\}$. The set of 2, 3 is a subset of the set of 1, 2, 3. The **universal set** is the general category set, or the set of all those elements under consideration. The **empty set,** or **null set,** is a set with no members—$\varnothing$ or { }.

Describing sets. Rule is a method of naming a set by describing its elements.

$$\{x \mid x > 3, x \text{ is a whole number}\}$$

$$\{\text{all students in the class with blue eyes}\}$$

Roster is a method of naming a set by listing its members.

$$\{4, 5, 6, \ldots\}$$

$$\{\text{Fred, Tom, Bob}\}$$

Venn diagrams (and **Euler circles**) are ways of pictorially describing sets, as shown in Figure 3.

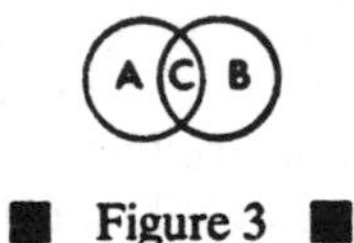

■ Figure 3 ■

Types of sets. Finite sets are countable; they stop—$\{1, 2, 3\} = \{3, 2, 1\}$. **Infinite sets** are uncountable; they continue forever—$\{1, 2, 3, \ldots\}$.

Comparing sets. Equal sets are those that have the exact same members—$\{1, 2, 3\} = \{3, 2, 1\}$. **Equivalent sets** are sets that have the same number of members—$\{1, 2, 3\} \sim \{a, b, c\}$.

Operations with sets. The **union** of two or more sets is a set containing all of the members in those sets.

Example 1: Find the union.

$$\{1, 2, 3\} \cup \{3, 4, 5\} = \{1, 2, 3, 4, 5\}$$

The union of sets with members 1, 2, 3 and 3, 4, 5 is the set with members 1, 2, 3, 4, 5.

The **intersection** of two or more sets is a set containing only the members contained in every set.

Example 2: Find the intersection.

$$\{1, 2, 3\} \cap \{3, 4, 5\} = \{3\}$$

The intersection of a set with members 1, 2, 3 and a set with members 3, 4, 5 is a set with only member 3.

Variables and Algebraic Expressions

A **variable** is a symbol used to denote any element of a given set—often a letter used to stand for a number. Variables are used to change verbal expressions into **algebraic expressions**.

Example 3: Give the algebraic expression.

	Verbal Expression	Algebraic Expression
(a)	the sum of a number and 7	$n + 7$ or $7 + n$
(b)	the number diminished by 10	$n - 10$
(c)	seven times a number	$7n$
(d)	x divided by 4	$x/4$
(e)	five more than the product of 2 and n	$2n + 5$ or $5 + 2n$

- **Key words denoting addition:**

sum	larger than	enlarge
plus	gain	rise
more than	increase	grow
greater than		

- **Key words denoting subtraction:**

difference	smaller than	lower
minus	fewer than	diminish
lose	decrease	reduced
less than	drop	

- **Key words denoting multiplication:**

product	times	of
multiplied by	twice	

- **Key words denoting division:**

quotient	ratio
divided by	half

Evaluating Expressions

To *evaluate* an *expression*, just replace the unknowns with grouping symbols, insert the value for the unknowns, and do the arithmetic.

Example 4: Evaluate each of the following.

(a) $ab + c$ if $a = 5, b = 4,$ and $c = 3$

$$5(4) + 3 =$$

$$20 + 3 = 23$$

(b) $2x^2 + 3y + 6$ if $x = 2$ and $y = 9$

$$2(2)^2 + 3(9) + 6 =$$
$$2(4) + 27 + 6 =$$
$$8 + 27 + 6 =$$
$$35 + 6 = 41$$

(c) $-4p^2 + 5q - 7$ if $p = -3$ and $q = -8$

$$-4(-3)^2 + 5(-8) - 7 =$$
$$-4(9) + 5(-8) - 7 =$$
$$-36 - 40 - 7 =$$
$$-76 - 7 = -83$$

(d) $\dfrac{a + c}{5} + \dfrac{a}{b + c}$ if $a = 3, b = -2$, and $c = 7$

$$\frac{(3) + (7)}{5} + \frac{(3)}{(-2) + (7)} =$$
$$\frac{10}{5} + \frac{3}{5} =$$
$$\frac{13}{5} = 2\tfrac{3}{5}$$

(e) $5x^3y^2$ if $x = -2$ and $y = 3$

$$5(-2)^3(3)^2 =$$
$$5(-8)(9) =$$
$$-40(9) = -360$$

Equations

An **equation** is a mathematical sentence, a relationship between numbers and/or symbols.

Axioms of equality. For all real numbers a, b, and c, the following are some basic rules for using the equal sign.

- **Reflexive axiom:** $a = a$.

  Therefore, $4 = 4$.

- **Symmetric axiom:** If $a = b$, then $b = a$.

  Therefore, if $2 + 3 = 5$, then $5 = 2 + 3$.

- **Transitive axiom:** If $a = b$ and $b = c$, then $a = c$

  Therefore, if $1 + 3 = 4$ and $4 = 2 + 2$, then $1 + 3 = 2 + 2$.

- **Additive axiom:** If $a = b$ and $c = d$, then $a + c = b + d$.

  Therefore, if $1 + 1 = 2$ and $3 + 3 = 6$,
  then $1 + 1 + 3 + 3 = 2 + 6$.

- **Multiplicative axiom:** If $a = b$ and $c = d$, then $ac = bd$.

  Therefore, if $1 = 2/2$ and $4 = 8/2$, then $1(4) = (2/2)(8/2)$.

Solving equations. Remember that an equation is like a balance scale, with the equal sign (=) being the fulcrum, or center. Thus, if you do the *same thing to both sides* of the equal sign (say, add 5 to each side), the equation will still be balanced.

Example 1: Solve for x.

$$x - 5 = 23$$

To solve the equation $x - 5 = 23$, you must get x by itself on one side; therefore, add 5 to both sides.

$$\begin{array}{rcr} x - 5 & = & 23 \\ +5 & & +5 \\ \hline x & = & 28 \end{array}$$

In the same manner, you may subtract, multiply, or divide *both* sides of an equation by the same (nonzero) number, and the equation will not change. Sometimes you may have to use more than one step to solve for an unknown.

Example 2: Solve for x.

$$3x + 4 = 19$$

Subtract 4 from both sides to get the $3x$ by itself on one side.

$$\begin{array}{rcr} 3x + 4 & = & 19 \\ -4 & & -4 \\ \hline 3x & = & 15 \end{array}$$

Then divide both sides by 3 to get x.

$$\frac{3x}{3} = \frac{15}{3}$$

$$x = 5$$

Remember that solving an equation is using opposite operations until the letter is on a side by itself (for addition, subtract; for multiplication, divide, etc.).

To check, substitute your answer into the original equation.

$$3x + 4 = 19$$

$$3(5) + 4 = 19$$

$$15 + 4 = 19$$

$$19 \overset{\checkmark}{=} 19$$

Example 3: Solve for x.

$$\frac{x}{5} - 4 = 2$$

Add 4 to both sides.

$$\begin{array}{rcr} \frac{x}{5} - 4 & = & 2 \\ +4 & & +4 \\ \hline \frac{x}{5} \quad & = & 6 \end{array}$$

Multiply both sides by 5 to get x.

$$(5)\frac{x}{5} = (5)6$$

$$x = 30$$

Example 4: Solve for x.

$$\frac{3}{5}x - 6 = 12$$

Add 6 to each side.

$$\begin{array}{r} \frac{3}{5}x - 6 = 12 \\ +6 \quad +6 \\ \hline \frac{3}{5}x \qquad = 18 \end{array}$$

Multiply each side by 5/3 (same as dividing by 3/5).

$$\left(\frac{5}{3}\right)\frac{3}{5}x = \left(\frac{5}{3}\right)18$$

$$x = \left(\frac{5}{\cancel{3}_1}\right)\frac{\cancel{18}^6}{1}$$

$$x = 30$$

Example 5: Solve for *x*.

$$5x = 2x - 6$$

Add $-2x$ to each side.

$$\begin{array}{r} 5x = \quad 2x - 6 \\ -2x \quad -2x \qquad \\ \hline 3x = \qquad -6 \end{array}$$

Divide both sides by 3.

$$\frac{3x}{3} = \frac{-6}{3}$$

$$x = -2$$

Example 6: Solve for *x*.

$$6x + 3 = 4x + 5$$

Add -3 to each side.

$$\begin{array}{rcl} 6x + 3 & = & 4x + 5 \\ -3 & & -3 \\ \hline 6x & = & 4x + 2 \end{array}$$

Add $-4x$ to each side.

$$\begin{array}{rcl} 6x & = & 4x + 2 \\ -4x & & -4x \\ \hline 2x & = & 2 \end{array}$$

Divide each side by 2.

$$\frac{2x}{2} = \frac{2}{2}$$

$$x = 1$$

Literal equations. Literal equations have no numbers, only symbols (letters).

Example 7: Solve for Q.

$$QP - X = Y$$

First add X to both sides.

$$\begin{array}{rcl} QP - X & = & Y \\ +X & & +X \\ \hline QP & = & Y + X \end{array}$$

Then divide both sides by P.

$$\frac{QP}{P} = \frac{Y + X}{P}$$

$$Q = \frac{Y + X}{P}$$

Operations opposite to those in the original equation were used to isolate Q. (To remove the $-X$, a $+X$ was *added* to both sides of the equation. Since the problem has Q times P, both sides were *divided* by P.

Example 8: Solve for y.

$$\frac{y}{x} = c$$

Multiply both sides by x to get y alone.

$$(x)\frac{y}{x} = (x)c$$

$$y = xc$$

Example 9: Solve for x.

$$\frac{b}{x} = \frac{p}{q}$$

To solve this equation quickly, you cross multiply. To cross multiply,

1. bring the denominators up next to the opposite side numerators and
2. multiply

$$\frac{b}{x} = \frac{p}{q}$$

$$bq = px$$

Then divide both sides by p to get x alone.

$$\frac{bq}{p} = \frac{px}{p}$$

$$\frac{bq}{p} = x \text{ or } x = \frac{bq}{p}$$

Cross multiplying can be used only when the format is two fractions separated by an equal sign.

Be aware that cross multiplying is most effective only when the letter you are solving for is on the *bottom* (the denominator) of a fraction. If it is on top (the numerator), it is easier simply to clear the denominator under the unknown you're solving for.

Example 10: Solve for x.

$$\frac{x}{k} = \frac{p}{q}$$

Multiply both sides by k.

$$(k)\frac{x}{k} = (k)\frac{p}{q}$$

$$x = \frac{kp}{q}$$

In this problem, there is no need to cross multiply.

Ratios and Proportions

Ratios. A **ratio** is a method of comparing two or more numbers or variables. Ratios are written as $a{:}b$ or in working form, as a fraction.

$$a/b \quad \text{or} \quad \frac{a}{b}$$

is read "*a* is to *b*." Notice that whatever comes after the "to" goes second or at the bottom of the fraction.

Proportions. Proportions are written as two ratios (fractions) equal to each other.

Example 11: Solve this problem for *x*.

$$p \text{ is to } q \text{ as } x \text{ is to } y$$

First the proportion may be rewritten.

$$\frac{p}{q} = \frac{x}{y}$$

Now simply multiply each side by *y*.

$$(y)\frac{p}{q} = (y)\frac{x}{y}$$

$$\frac{yp}{q} = x$$

Example 12: Solve this proportion for *t*.

$$s \text{ is to } t \text{ as } r \text{ is to } q$$

Rewrite.

$$\frac{s}{t} = \frac{r}{q}$$

Cross multiply.

$$sq = rt$$

Divide both sides by r.

$$\frac{sq}{r} = \frac{rt}{r}$$

$$\frac{sq}{r} = t$$

Solving proportions for value. Follow the procedures given in Examples 11 and 12 to solve for the unknown.

Example 13: Solve for x.

$$\frac{4}{x} = \frac{2}{5}$$

Cross multiply.

$$(4)(5) = 2x$$

$$20 = 2x$$

Divide both sides by 2.

$$\frac{20}{2} = \frac{2x}{2}$$

$$10 = x$$

Solving Systems of Equations (Simultaneous Equations)

If you have two equations with the same two unknowns in each, you can solve for both unknowns. There are three common methods for solving: addition/subtraction, substitution, and graphing.

Addition/subtraction method. To use the **addition/subtraction method,**

1. Multiply one or both equations by some number to make the number in front of one of the letters (unknowns) the same in each equation.
2. Add or subtract the two equations to eliminate one letter.
3. Solve for the other unknown.
4. Insert the value of the first unknown in one of the original equations to solve for the second unknown.

Example 1: Solve for x and y.

$$3x + 3y = 24$$

$$2x + y = 13$$

First multiply the bottom equation by 3. Now the y is preceded by a 3 in each equation.

$$3x + 3y = 24 \qquad 3x + 3y = 24$$

$$3(2x) + 3(y) = 3(13) \qquad 6x + 3y = 39$$

Now the equations can be subtracted, eliminating the y terms.

$$\begin{array}{r} 3x + 3y = 24 \\ \underline{-6x + -3y = -39} \\ -3x = -15 \end{array}$$

$$\frac{-3x}{-3} = \frac{-15}{-3}$$

$$x = 5$$

Now insert $x = 5$ in one of the original equations to solve for y.

$$2x + y = 13$$

$$2(5) + y = 13$$

$$\begin{array}{r} 10 + y = 13 \\ \underline{-10 -10} \\ y = 3 \end{array}$$

Answer: $x = 5, y = 3$

Of course, if the number in front of a letter is already the same in each equation, you do not have to change either equation. Simply add or subtract.

Example 2: Solve for x and y.

$$x + y = 7$$

$$x - y = 3$$

$$\begin{array}{r} x + y = 7 \\ \underline{x - y = 3} \\ 2x = 10 \end{array}$$

$$\frac{2x}{2} = \frac{10}{2}$$

$$x = 5$$

Now, inserting 5 for x in the first equation gives

$$\begin{array}{r} 5 + y = \ \ 7 \\ \underline{-5 \qquad\ \ -5} \\ y = \ \ 2 \end{array}$$

Answer: $x = 5, y = 2$

You should note that this method will not work when the two equations are, in fact, the same.

Example 3: Solve for a and b.

$$3a + 4b = 2$$

$$6a + 8b = 4$$

The second equation is actually the first equation multiplied by 2. In this instance, the *system is unsolvable.*

Example 4: Solve for p and q.

$$3p + 4q = 9$$

$$2p + 2q = 6$$

Multiply the second equation by 2.

$$(2)2p + (2)2q = (2)6$$

$$4p + 4q = 12$$

Now subtract the equations.

$$\begin{array}{rcr} 3p + 4q &=& 9 \\ \underline{(-)4p + 4q} &\underline{=}& \underline{12} \\ -p &=& -3 \\ p &=& 3 \end{array}$$

Now that you know $p = 3$, you may plug in 3 for p in either of the two original equations to find q.

$$3p + 4q = 9$$

$$3(3) + 4q = 9$$

$$9 + 4q = 9$$

$$4q = 0$$

$$q = 0$$

Answer: $p = 3, q = 0$

Substitution method. Sometimes a system is more easily solved by the **substitution method.** This method involves substituting one equation into another.

Example 5: Solve for x and y.

$$x = y + 8$$

$$x + 3y = 48$$

From the first equation, substitute $(y + 8)$ for x in the second equation.

$$(y + 8) + 3y = 48$$

Now solve for y. Simplify by combining y's.

$$\begin{array}{rl} 4y + 8 &= 48 \\ -8 & -8 \\ \hline 4y \quad &= 40 \end{array}$$

$$\frac{4y}{4} = \frac{40}{4}$$

$$y = 10$$

Now insert $y = 10$ in one of the original equations.

$$x = y + 8$$

$$x = 10 + 8$$

$$x = 18$$

Answer: $y = 10, x = 18$

Graphing method. Another method of solving equations is by **graphing** each equation on a coordinate graph. The coordinates of the intersection will be the solution to the system. If you are unfamiliar with coordinate graphing, carefully review the chapter on analytic geometry (page 101) before attempting this method.

Example 6: Solve the system by graphing

$$x = 4 + y$$

$$x - 3y = 4$$

First, find three values for x and y that satisfy each equation. (Although only two points are necessary to determine a straight line, finding a third point is a good way of checking.)

$x = 4 + y$

x	y
4	0
2	−2
5	1

$x - 3y = 4$

x	y
1	−1
4	0
7	1

Now graph the two lines on the coordinate plane.

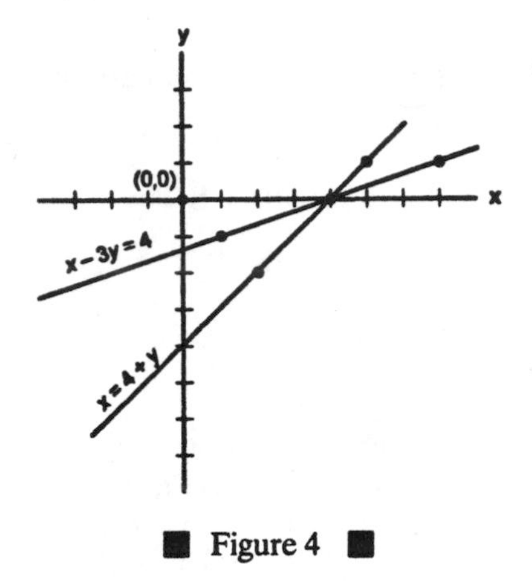

■ Figure 4 ■

As shown in Figure 4, the point where the two lines cross (4, 0) is the solution of the system.

If the lines are parallel, they do not intersect, and therefore, there is *no solution to that system.*

Monomials

A **monomial** is an algebraic expression that consists of only one term. (A term is a numerical or literal expression with its own sign.) For instance, $9x$, $4a^2$ and $3mpxz^2$ are all monomials. The number in front of the variable is called the **numerical coefficient**. In $9y$, 9 is the coefficient.

Adding and subtracting monomials. To *add* or *subtract monomials,* follow the same rules as with signed numbers (page 19), *provided that the terms are alike.* Notice that you add or subtract the coefficients only and leave the variables the same.

Example 1: Perform the operation indicated.

(a)
$$\begin{array}{r} 15x^2yz \\ -\ 18x^2yz \\ \hline -3x^2yz \end{array}$$

(b) $3x + 2x = 5x$

(c)
$$\begin{array}{r} 9y \\ -\ 3y \\ \hline 6y \end{array}$$

(d)
$$\begin{aligned} 17q + 8q - 3q - (-4q) &= \\ 22q - (-4q) &= \\ 22q + 4q &= 26q \end{aligned}$$

Remember that the rules for signed numbers apply to monomials as well.

Multiplying monomials. Reminder: The rules and definitions for powers and exponents (page 9) also apply in algebra.

$$5 \cdot 5 = 5^2 \quad \text{and} \quad x \cdot x = x^2$$

Similarly, $$a \cdot a \cdot a \cdot b \cdot b = a^3b^2$$

To *multiply monomials,* add the exponents of the same bases.

Example 2: Multiply the following.

(a) $(x^3)(x^4) = x^7$

(b) $(x^2y)(x^3y^2) = x^5y^3$

(c) $(6k^5)(5k^2) = 30k^7$ (multiply numbers)

(d) $-4(m^2n)(-3m^4n^3) = 12m^6n^4$

(e) $(c^2)(c^3)(c^4) = c^9$

(f) $(3a^2b^3c)(b^2c^2d) = 3a^2b^5c^3d$

Note that in example (d) the product of -4 and -3 is $+12$, the product of m^2 and m^4 is m^6, and the product of n and n^3 is n^4, since any monomial having no exponent indicated is assumed to have an exponent of 1.

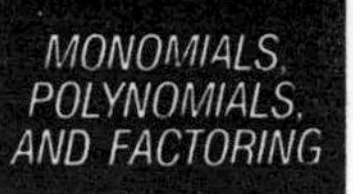

When monomials are being *raised to a power,* the answer is obtained by multiplying the exponents of each part of the monomial by the power to which it is being raised.

Example 3: Simplify.

(a) $(a^7)^3 = a^{21}$

(b) $(x^3y^2)^4 = x^{12}y^8$

(c) $(2x^2y^3)^3 = (2)^3x^6y^9 = 8x^6y^9$

Dividing monomials. To *divide monomials,* subtract the exponent of the divisor from the exponent of the dividend of the same base.

Example 4: Divide.

(a) $\dfrac{y^{15}}{y^4} = y^{11}$ or $y^{15} \div y^4 = y^{11}$

(b) $\dfrac{x^5y^2}{x^3y} = x^2y$

(c) $\dfrac{36a^4b^6}{-9ab} = -4a^3b^5$ (divide the numbers)

(d) $\dfrac{fg^{15}}{g^3} = fg^{12}$

(e) $\dfrac{x^5}{x^8} = \dfrac{1}{x^3}$ (may also be expressed x^{-3})

(f) $\dfrac{-3(xy)(xy^2)}{xy}$

You can simplify the numerator first.

$$\frac{-3(xy)(xy^2)}{xy} = \frac{-3x^2y^3}{xy} = -3xy^2$$

Or, since the numerator is all multiplication, you can cancel.

$$\frac{-3(\cancel{xy})(xy^2)}{\cancel{xy}} = -3xy^2$$

Working with negative exponents. Remember, if the exponent is negative, such as x^{-3}, then the variable and exponent may be dropped under the number 1 in a fraction to remove the negative sign as follows.

$$x^{-3} = \frac{1}{x^3}$$

Example 5: Express the answers with positive exponents.

(a) $a^{-2}b = \dfrac{b}{a^2}$

(b) $\dfrac{a^{-3}}{b^4} = \dfrac{1}{a^3b^4}$

(c) $(a^2b^{-3})(a^{-1}b^4) = ab$

$$\begin{bmatrix} a^2 \cdot a^{-1} = a \\ b^{-3} \cdot b^4 = b \end{bmatrix}$$

Polynomials

A **polynomial** consists of two or more terms. For example, $x + y$, $y^2 - x^2$, and $x^2 + 3x + 5y^2$ are all polynomials. A **binomial** is a polynomial that consists of exactly two terms. For example, $x + y$ is a binomial. A **trinomial** is a polynomial that consists of exactly three terms. For example, $y^2 + 9y + 8$ is a trinomial.

Polynomials are usually arranged in one of two ways. **Ascending order** is basically when the power of a term increases for each succeeding term. For example, $x + x^2 + x^3$ or $5x + 2x^2 - 3x^3 + x^5$ are arranged in ascending order. **Descending order** is basically when the power of a term decreases for each succeeding term. For example, $x^3 + x^2 + x$ or $2x^4 + 3x^2 + 7x$ are arranged in descending order. Descending order is more commonly used.

Adding and subtracting polynomials. To *add* or *subtract polynomials,* just arrange *like terms* in columns and then add or subtract. (Or simply add or subtract like terms when rearrangement is not necessary.)

Example 6: Add or subtract as indicated.

(a) Add.

$$\begin{array}{r} a^2 + \ ab + \ b^2 \\ 3a^2 + 4ab - 2b^2 \\ \hline 4a^2 + 5ab - \ b^2 \end{array}$$

(b) $(5y - 3x) + (9y + 4x) =$

$(5y - 3x) + (9y + 4x) = 14y + x$ or $x + 14y$

(c) Subtract.

$$\begin{array}{r} a^2 + b^2 \\ (-)\ 2a^2 - b^2 \\ \hline \end{array} \qquad \begin{array}{r} a^2 + b^2 \\ (+) - 2a^2 + b^2 \\ \hline -a^2 + 2b^2 \end{array}$$

(d) $(3cd - 6mt) - (2cd - 4mt) =$

$(3cd - 6mt) + (-2cd + 4mt) =$

$(3cd - 6mt) + (-2cd + 4mt) = cd - 2mt$

(e) $3a^2bc + 2ab^2c + 4a^2bc + 5ab^2c =$

$$\begin{array}{r} 3a^2bc + 2ab^2c \\ + 4a^2bc + 5ab^2c \\ \hline 7a^2bc + 7ab^2c \end{array}$$

or

$3a^2bc + 2ab^2c + 4a^2bc + 5ab^2c = 7a^2bc + 7ab^2c$

Multiplying polynomials. To *multiply polynomials,* multiply each term in one polynomial by each term in the other polynomial. Then simplify if necessary.

Example 7: Multiply.

$$\begin{array}{r} 2x - 2a \\ \times \quad 3x + a \\ \hline + 2ax - 2a^2 \\ 6x^2 - 6ax \quad\quad\; \\ \hline 6x^2 - 4ax - 2a^2 \end{array} \quad \text{similar to} \quad \begin{array}{r} 21 \\ \times 23 \\ \hline 63 \\ 42\; \\ \hline 483 \end{array}$$

Or you may wish to use the "F.O.I.L." method with *binomials*. F.O.I.L. means **F**irst terms, **O**utside terms, **I**nside terms, **L**ast terms. Then simplify if necessary.

Example 8: Multiply.

$$(3x + a)(2x - 2a) =$$

Multiply *first* terms from each quantity.

$$(3x + a)(2x - 2a) = 6x^2 \_\_\_\_\_\_\_\_\_\_\_\_$$

Now *outside* terms.

$$(3x + a)(2x - 2a) = 6x^2 - 6ax \_\_\_\_\_\_\_\_$$

Now *inside* terms.

$$(3x + a)(2x - 2a) = 6x^2 - 6ax + 2ax \_\_\_\_$$

Finally *last* terms.

$$(3x + a)(2x - 2a) = 6x^2 - 6ax + 2ax - 2a^2$$

Now simplify.

$$6x^2 - 6ax + 2ax - 2a^2 = 6x^2 - 4ax - 2a^2$$

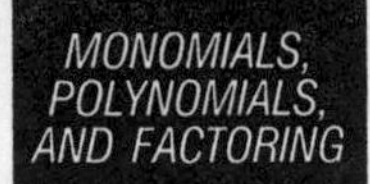

Example 9: Multiply.

$$(x + y)(x + y + z) =$$

$$\begin{array}{r} x + y + z \\ \times \qquad\qquad x + y \\ \hline xy + y^2 + yz \\ x^2 + xz + xy \qquad\qquad \\ \hline x^2 + xz + 2xy + y^2 + yz \end{array}$$

This operation can also be done using the distributive property.

$$(x + y)(x + y + z) = x^2 + xy + xz + xy + y^2 + yz$$

$$= x^2 + 2xy + xz + yz + y^2$$

Dividing polynomials by monomials. To *divide a polynomial by a monomial,* just divide each term in the polynomial by the monomial.

Example 10: Divide.

(a) $(6x^2 + 2x) \div 2x =$

$$\frac{6x^2 + 2x}{2x} =$$

$$\frac{6x^2}{2x} + \frac{2x}{2x} = 3x + 1$$

(b) $(16a^7 - 12a^5) \div 4a^2 =$

$$\frac{16a^7 - 12a^5}{4a^2} =$$

$$\frac{16a^7}{4a^2} - \frac{12a^5}{4a^2} = 4a^5 - 3a^3$$

Dividing polynomials by polynomials. To *divide a polynomial by a polynomial,* make sure both are in descending order; then use long division. (Remember: Divide by the first term, multiply, subtract, bring down.)

Example 11: Divide $4a^2 + 18a + 8$ by $a + 4$.

First divide a into $4a^2$

$$\begin{array}{r} 4a \\ a+4 \overline{)4a^2 + 18a + 8} \end{array}$$

Now multiply $4a$ times $(a + 4)$

$$\begin{array}{r} 4a \\ a+4 \overline{)4a^2 + 18a + 8} \\ \underline{4a^2 + 16a} \end{array}$$

Now subtract.

$$\begin{array}{r} 4a \\ a+4 \overline{)4a^2 + 18a + 8} \\ (-)\ \underline{4a^2 + 16a} \\ 2a \end{array}$$

Now bring down the +8.

$$\begin{array}{r} 4a \\ a+4 \overline{)4a^2 + 18a + 8} \\ (-)\ \underline{4a^2 + 16a} \\ 2a + 8 \end{array}$$

Now divide a into $2a$.

$$\begin{array}{r} 4a + 2 \\ a+4 \overline{)4a^2 + 18a + 8} \\ (-)\ \underline{4a^2 + 16a} \\ 2a + 8 \end{array}$$

Now multiply 2 times $(a + 4)$.

$$\begin{array}{r} 4a + 2 \\ a+4 \overline{)4a^2 + 18a + 8} \\ (-)\ \underline{4a^2 + 16a} \\ 2a + 8 \\ \underline{2a + 8} \end{array}$$

Now subtract.

$$\begin{array}{r} 4a + 2 \\ a + 4\,\overline{)4a^2 + 18a + 8} \\ (-)\ \underline{4a^2 + 16a} \quad \\ 2a + 8 \\ (-)\ \underline{2a + 8} \\ 0 \end{array}$$

$$\begin{array}{r} 4a + 2 \\ a + 4\,\overline{)4a^2 + 18a + 8} \\ (-)\ \underline{4a^2 + 16a} \quad \\ 2a + 8 \\ (-)\ \underline{2a + 8} \\ 0 \end{array} \quad \textit{similar to} \quad \begin{array}{r} 23 \\ 53\,\overline{)1219} \\ (-)\ \underline{106}\ \ \\ 159 \\ (-)\ \underline{159} \\ 0 \end{array}$$

Example 12: Divide.

(a) $(3x^2 + 4x + 1) \div (x + 1) =$

$$\begin{array}{r} 3x + 1 \\ x + 1\,\overline{)3x^2 + 4x + 1} \\ (-)\ \underline{3x^2 + 3x} \quad \\ x + 1 \\ (-)\ \underline{x + 1} \\ 0 \end{array}$$

(b) $(2x + 1 + x^2) \div (x + 1) =$

First change to descending order: $x^2 + 2x + 1$. Then divide.

$$\begin{array}{r} x + 1 \\ x + 1\,\overline{)x^2 + 2x + 1} \\ (-)\ \underline{x^2 + 1x} \quad \\ x + 1 \\ (-)\ \underline{x + 1} \\ 0 \end{array}$$

(c) $(m^3 - m) \div (m + 1) =$

Note: When terms are missing, be sure to leave proper room between terms.

$$
\begin{array}{r}
m^2 - m \\
m + 1 \,\overline{)\, m^3 + 0m^2 - m} \\
(-)\, \underline{m^3 + m^2 } \\
- m^2 - m \\
(-)\, \underline{- m^2 - m} \\
0
\end{array}
$$

(d) $(10a^2 - 29a - 21) \div (2a - 7) =$

$$
\begin{array}{r}
5a + 3 \\
2a - 7 \,\overline{)\, 10a^2 - 29a - 21} \\
(-)\, \underline{10a^2 - 35a } \\
6a - 21 \\
(-)\, \underline{6a - 21} \\
0
\end{array}
$$

(e) $(x^2 + 2x + 4) \div (x + 1) =$

Note that remainders are possible.

$$
\begin{array}{r}
x + 1 \text{ (with remainder 3)} \\
x + 1 \,\overline{)\, x^2 - 2x + 4} \\
(-)\, \underline{x^2 + x } \\
x + 4 \\
(-)\, \underline{x + 1} \\
3
\end{array}
$$

This answer can be rewritten as $(x + 1) + \dfrac{3}{x + 1}$.

Factoring

To **factor** means to find two or more quantities whose product equals the original quantity.

Factoring out a common factor. To *factor out a common factor,* (1) find the largest common monomial factor of each term and (2) divide the original polynomial by this factor to obtain the second factor. The second factor will be a polynomial.

Example 13: Factor.

(a) $5x^2 + 4x = x(5x + 4)$

(b) $2y^3 - 6y = 2y(y^2 - 3)$

(c) $x^5 - 4x^3 + x^2 = x^2(x^3 - 4x + 1)$

When the common monomial factor is the last term, 1 is used as a place holder in the second factor.

Factoring the difference between two squares. To *factor the difference between two squares,* (1) find the square root of the first term and the square root of the second term and (2) express your answer as the product of the sum of the quantities from step 1 times the difference of those quantities.

Example 14: Factor.

(a) $x^2 - 144 = (x + 12)(x - 12)$

Note: $x^2 + 144$ is *not* factorable.

(b) $a^2 - b^2 = (a + b)(a - b)$

(c) $9y^2 - 1 = (3y + 1)(3y - 1)$

Factoring polynomials having three terms of the form $ax^2 + bx + c$. To *factor polynomials having three terms of the form* $ax^2 + bx + c$, (1) check to see if you can monomial factor (factor out common terms). Then if $a = 1$ (that is, the first term is simply x^2), use double parentheses and factor the first term. Place these factors in the left sides of the parentheses. For example,

$$(x \quad)(x \quad)$$

(2) Factor the last term and place the factors in the right sides of the parentheses.

To decide on the signs of the numbers, do the following. If the sign of the last term is *negative*, (1) find two numbers (one will be a positive number and the other a negative number) whose product is the last term and whose *difference* is the *coefficient* (number in front) of the middle term and (2) give the larger of these two numbers the sign of the middle term and the *opposite* sign to the other factor.

If the sign of the last term is *positive,* (1) find two numbers (both will be positive or both will be negative) whose product is the last term and whose *sum* is the coefficient of the middle term and (2) give both factors the sign of the middle term.

Example 15: Factor $x^2 - 3x - 10$.

First check to see if you can monomial factor (factor out common terms). Since this is not possible, use double parentheses and factor the first term as follows: $(x \quad)(x \quad)$. Next, factor the last term, 10, into 2 times 5 (5 must take the negative sign and 2 must take the positive sign because they will then total the coefficient of the middle term, which is −3) and add the proper signs, leaving

$$(x - 5)(x + 2)$$

Multiply **means** (inner terms) and **extremes** (outer terms) to check.

$$(x - 5)(x + 2)$$

$$\begin{array}{l} -5x \\ \underline{+2x} \\ -3x \text{ (which is the middle term)} \end{array}$$

To completely check, multiply the factors together.

$$\begin{array}{r} x - 5 \\ \underline{\times \quad x + 2} \\ + 2x - 10 \\ \underline{x^2 - 5x \qquad} \\ x^2 - 3x + 10 \end{array}$$

Example 16: Factor $x^2 + 8x + 15$.

$$(x + 3)(x + 5)$$

Notice that $3 \times 5 = 15$ and $3 + 5 = 8$, the coefficient of the middle term. Also note that the signs of both factors are +, the sign of the middle term. To check,

$$(x + 3)(x + 5)$$

$$\begin{array}{l} + 3x \\ \underline{+ 5x} \\ + 8x \text{ (the middle term)} \end{array}$$

Example 17: Factor $x^2 - 5x - 14$.

$$(x - 7)(x + 2)$$

Notice that $7 \times 2 = 14$ and $7 - 2 = 5$, the coefficient of the middle term. Also note that the sign of the larger factor, 7, is −, while the other factor, 2, has a + sign. To check,

$$(x - 7)(x + 2)$$

$$\begin{array}{l} -7x \\ \underline{+2x} \\ -5x \text{ (the middle term)} \end{array}$$

If, however, $a \neq 1$ (that is, the first term has a coefficient—for example, $4x^2 + 5x + 1$), then additional trial and error will be necessary.

Example 18: Factor $4x^2 + 5x + 1$.

$(2x + \quad)(2x + \quad)$ might work for the first term. But when 1's are used as factors to get the last term—$(2x + 1)(2x + 1)$—the middle term comes out as $4x$ instead of $5x$.

$$(2x + 1)(2x + 1)$$

$$\begin{array}{l} +2x \\ \underline{+2x} \\ +4x \end{array}$$

Therefore, try $(4x + \quad)(x + \quad)$. Now using 1's as factors to get the last terms gives $(4x + 1)(x + 1)$. Checking for the middle term,

$$(4x + 1)(x + 1)$$

$$\begin{array}{r} +\,1x \\ \underline{+\,4x} \\ +\,5x \end{array}$$

Therefore, $4x^2 + 5x + 1 = (4x + 1)(x + 1)$.

Example 19: Factor $4a^2 + 6a + 2$.

Factoring out a 2 leaves

$$2(2a^2 + 3a + 1)$$

Now factor as usual, giving

$$2(2a + 1)(a + 1)$$

To check,

$$(2a + 1)(a + 1)$$

$$\begin{array}{r} +\,1a \\ \underline{+\,2a} \\ +\,3a \end{array}$$

(the middle term after 2 was factored out)

Example 20: Factor $5x^3 + 6x^2 + x$.

Factoring out an x leaves

$$x(5x^2 + 6x + 1)$$

Now factor as usual, giving

$$x(5x + 1)(x + 1)$$

To check,

$$(5x + 1)(x + 1)$$

$+1x$

$+5x$ (the middle term after

$+6x$ x was factored out)

Example 21: Factor $5 + 7b + 2b^2$ (a slight twist).

$$(5 + 2b)(1 + b)$$

To check,

$$(5 + 2b)(1 + b)$$

$+2b$

$+5b$

$+7b$ (the middle term)

Note that $(5 + b)(1 + 2b)$ is incorrect because it gives the wrong middle term.

Example 22: Factor $x^2 + 2xy + y^2$.

$$(x + y)(x + y)$$

To check,

$$(x + y)(x + y)$$

$$+xy$$

$$+xy$$

$$+2xy \quad \text{(the middle term)}$$

Note: There are polynomials that are *not* factorable.

Factoring by grouping. Some polynomials have binomial, trinomial, and other polynomial factors.

Example 23: Factor $x + 2 + xy + 2y$.

Since there is no monomial factor, you should attempt rearranging the terms and looking for binomial factors.

$$x + 2 + xy + 2y = x + xy + 2 + 2y$$

Now grouping gives

$$(x + xy) + (2 + 2y)$$

Now factoring gives

$$x(1 + y) + 2(1 + y)$$

Using the distributive property gives

$$(x + 2)(1 + y)$$

You could rearrange them differently, but you would still come up with the same factoring.

Algebraic fractions are fractions using a variable in the numerator or denominator, such as $3/x$. Since division by 0 is impossible, variables in the denominator have certain restrictions. The denominator can *never* equal 0. Therefore, in the fractions

$\frac{5}{x}$ x cannot equal 0 ($x \neq 0$)

$\frac{2}{x-3}$ x cannot equal 3 ($x \neq 3$)

$\frac{3}{a-b}$ $a - b$ cannot equal 0 ($a - b \neq 0$) so a cannot equal b ($a \neq b$)

$\frac{4}{a^2b}$ a cannot equal 0 and b cannot equal 0 ($a \neq 0$ and $b \neq 0$)

Be aware of these types of restrictions.

Operations with Algebraic Fractions

Reducing algebraic fractions. To *reduce an algebraic fraction* to lowest terms, first factor the numerator and the denominator; then cancel (or divide out) common factors.

Example 1: Reduce.

(a) $\frac{4x^3}{8x^2} = \frac{\overset{1}{\cancel{4}}x^{\overset{1}{\cancel{3}}}}{\underset{2}{\cancel{8}}\cancel{x^2}} = \frac{1}{2}x$ or $\frac{x}{2}$

(b) $\frac{3x-3}{4x-4}=\frac{3(x-1)}{4(x-1)}=\frac{3\cancel{(x-1)}}{4\cancel{(x-1)}}=\frac{3}{4}$

(c) $\frac{x^2+2x+1}{3x+3}=\frac{(x+1)(x+1)}{3(x+1)}=\frac{\cancel{(x+1)}(x+1)}{3\cancel{(x+1)}}=\frac{x+1}{3}$

Warning: Do not cancel through an addition or subtraction sign as shown here.

$$\frac{x+1}{x+2}\neq\frac{\cancel{x}+1}{\cancel{x}+2}\neq\frac{1}{2}$$

or

$$\frac{x+6}{6}\neq\frac{x+\cancel{6}}{\cancel{6}}\neq x$$

Multiplying algebraic fractions. To *multiply algebraic fractions,* first factor the numerators and denominators that are polynomials; then cancel where possible. Multiply the remaining numerators together and denominators together. (If you've canceled properly, your answer will be in reduced form.)

Example 2: Multiply.

(a) $\frac{2x}{3}\cdot\frac{y}{5}=\frac{2x\times y}{3\times 5}=\frac{2xy}{15}$

(b) $\frac{x^2}{3y}\cdot\frac{2y}{3x}=\frac{\cancel{x^2}^{1}}{3\cancel{y}}\cdot\frac{2\cancel{y}}{3\cancel{x}}=\frac{2x}{9}$

(c) $\dfrac{x+1}{5y+10} \cdot \dfrac{y+2}{x^2+2x+1} = \dfrac{x+1}{5(y+2)} \cdot \dfrac{y+2}{(x+1)(x+1)} =$

$\dfrac{\cancel{x+1}^{1}}{5(\cancel{y+2})} \cdot \dfrac{\cancel{y+2}^{1}}{(\cancel{x+1})(x+1)} = \dfrac{1}{5(x+1)}$

(d) $\dfrac{x^2-4}{6} \cdot \dfrac{3y}{2x+4} = \dfrac{(x+2)(x-2)}{6} \cdot \dfrac{3y}{2(x+2)} =$

$\dfrac{(\cancel{x+2})(x-2)}{{}_2\cancel{6}} \cdot \dfrac{\cancel{3}^{1}y}{2(\cancel{x+2})} = \dfrac{(x-2)y}{4}$

(e) $\dfrac{x^2+4x+4}{x-3} \cdot \dfrac{5}{3x+6} = \dfrac{(x+2)(x+2)}{x-3} \cdot \dfrac{5}{3(x+2)} =$

$\dfrac{(x+2)(\cancel{x+2})}{x-3} \cdot \dfrac{5}{3(\cancel{x+2})} = \dfrac{5(x+2)}{3(x-3)}$

Dividing algebraic fractions. To *divide algebraic fractions,* invert the fraction and multiply. Remember, you can cancel only after you invert.

Example 3: Divide.

(a) $\dfrac{3x^2}{5} \div \dfrac{2x}{y} = \dfrac{3x^2}{5} \cdot \dfrac{y}{2x} = \dfrac{3x^{\cancel{2}\,1}}{5} \cdot \dfrac{y}{2\cancel{x}} = \dfrac{3xy}{10}$

(b) $\dfrac{4x-8}{6} \div \dfrac{x-2}{3} = \dfrac{4x-8}{6} \cdot \dfrac{3}{x-2} = \dfrac{4(x-2)}{6} \cdot \dfrac{3}{x-2} =$

$\dfrac{4(\cancel{x-2})^{1}}{{}_2\cancel{6}} \cdot \dfrac{\cancel{3}^{1}}{\underset{1}{\cancel{x-2}}} = \dfrac{4}{2} = 2$

Adding or subtracting algebraic fractions. To *add* or *subtract* *algebraic fractions having a common denominator,* simply keep the denominator and combine (add or subtract) the numerators. Reduce if possible.

Example 4: Perform the indicated operation.

(a) $\frac{4}{x} + \frac{5}{x} = \frac{4+5}{x} = \frac{9}{x}$

(b) $\frac{x-4}{x+1} + \frac{3}{x+1} = \frac{x-4+3}{x+1} = \frac{x-1}{x+1}$

(c) $\frac{3x}{y} - \frac{2x-1}{y} = \frac{3x-(2x-1)}{y} = \frac{3x-2x+1}{y} = \frac{x+1}{y}$

To *add* or *subtract algebraic fractions having different denominators,* first find a lowest common denominator (LCD), change each fraction to an equivalent fraction with the common denominator, then combine each numerator. Reduce if possible.

Example 5: Perform the indicated operation.

(a) $\frac{2}{x} + \frac{3}{y} =$

$\text{LCD} = xy$

$\frac{2}{x} \cdot \frac{y}{y} + \frac{3}{y} \cdot \frac{x}{x} = \frac{2y}{xy} + \frac{3x}{xy} = \frac{2y+3x}{xy}$

(b) $\frac{x+2}{3x} + \frac{x-3}{6x} =$

LCD $= 6x$

$$\frac{x+2}{3x} \cdot \frac{2}{2} + \frac{x-3}{6x} = \frac{2x+4}{6x} + \frac{x-3}{6x} =$$

$$\frac{2x+4+x-3}{6x} = \frac{3x+1}{6x}$$

If there is a common variable factor with more than one exponent, use its greatest exponent.

Example 6: Perform the indicated operation.

(a) $\frac{2}{y^2} - \frac{3}{y} =$

LCD $= y^2$

$$\frac{2}{y^2} - \frac{3}{y} \cdot \frac{y}{y} = \frac{2}{y^2} - \frac{3y}{y^2} = \frac{2-3y}{y^2}$$

(b) $\frac{4}{x^3y} + \frac{3}{xy^2} =$

LCD $= x^3y^2$

$$\frac{4}{x^3y} \cdot \frac{y}{y} + \frac{3}{xy^2} \cdot \frac{x^2}{x^2} = \frac{4y}{x^3y^2} + \frac{3x^2}{x^3y^2} = \frac{4y+3x^2}{x^3y^2}$$

(c) $\frac{x}{x+1} - \frac{2x}{x+2} =$

$\text{LCD} = (x+1)(x+2)$

$$\frac{x}{x+1} \cdot \frac{(x+2)}{(x+2)} - \frac{2x}{x+2} \cdot \frac{(x+1)}{(x+1)} =$$

$$\frac{x^2+2x}{(x+1)(x+2)} - \frac{2x^2+2x}{(x+1)(x+2)} =$$

$$\frac{x^2+2x-2x^2-2x}{(x+1)(x+2)} = \frac{-x^2}{(x+1)(x+2)}$$

To find the lowest common denominator, it is often necessary to factor the denominators and proceed as follows.

Example 7: Perform the indicated operation.

$$\frac{2x}{x^2-9} - \frac{5}{x^2+4x+3} = \frac{2x}{(x+3)(x-3)} - \frac{5}{(x+3)(x+1)} =$$

$\text{LCD} = (x+3)(x-3)(x+1)$

$$\frac{2x}{(x+3)(x-3)} \cdot \frac{(x+1)}{(x+1)} - \frac{5}{(x+3)(x+1)} \cdot \frac{(x-3)}{(x-3)} =$$

$$\frac{2x^2+2x}{(x+3)(x-3)(x+1)} - \frac{5x-15}{(x+3)(x-3)(x+1)} =$$

$$\frac{2x^2+2x-(5x-15)}{(x+3)(x-3)(x+1)} = \frac{2x^2+2x-5x+15}{(x+3)(x-3)(x+1)} =$$

$$\frac{2x^2-3x+15}{(x+3)(x-3)(x+1)}$$

Inequalities

An **inequality** is a statement in which the relationships are not equal. Instead of using an equal sign (=) as in an equation, these symbols are used: > (greater than) and < (less than) or ≥ (greater than or equal to) and ≤ (less than or equal to).

Axioms and properties of inequalities. For all real numbers a, b, and c, the following are some basic rules for using the inequality signs.

- **Trichotomy axiom:** $a > b$, $a = b$, or $a < b$.

  These are the only possible relationships between two numbers. Either the first number is greater than the second, the numbers are equal, or the first number is less than the second.

- **Transitive axiom:** If $a > b$, and $b > c$, then $a > c$.

  Therefore, if $3 > 2$ and $2 > 1$, then $3 > 1$.

  If $a < b$ and $b < c$, then $a < c$.

  Therefore, if $4 < 5$ and $5 < 6$, then $4 < 6$.

- **Addition property:** If $a > b$, then $a + c > b + c$.

  Therefore, if $3 > 2$, then $3 + 1 > 2 + 1$.

- **Positive multiplication property:** If $c > 0$, then $a > b$ if, and only if, $ac > bc$.

  Therefore, if $2 > 0$, then $3 > 1$ if, and only if, $3(2) > 1(2)$.

- **Negative multiplication property:** If $c < 0$, then $a > b$ if, and only if, $ac < bc$.

  Therefore, if $-2 < 0$,

  then $5 > 3$ if, and only if, $5(-2) < 3(-2)$.

  Reverse the inequality sign when multiplying (or dividing) by a negative number.

Solving inequalities. When working with inequalities, treat them exactly like equations (*except*, if you multiply or divide both sides by a negative number, you must *reverse* the direction of the sign).

Example 1: Solve for x: $2x + 4 > 6$.

$$\begin{array}{rcr} 2x + 4 & > & 6 \\ -4 & & -4 \\ \hline 2x & > & 2 \end{array}$$

$$\frac{2x}{2} > \frac{2}{2}$$

$$x > 1$$

Answers are sometimes written in set builder notation $\{x : x > 1\}$, which is read "the set of all x such that x is greater than 1."

Example 2: Solve for x: $-7x > 14$.

Divide by -7 and reverse the sign.

$$\frac{-7x}{-7} < \frac{14}{-7}$$

$$x < -2$$

Example 3: Solve for x: $3x + 2 \geq 5x - 10$.

$$\begin{array}{rcr} 3x + 2 & \geq & 5x - 10 \\ -2 & & -\ 2 \\ \hline 3x & \geq & 5x - 12 \end{array}$$

$$\begin{array}{rcr} 3x & \geq & 5x - 12 \\ -5x & & -5x \\ \hline -2x & \geq & -\ 12 \end{array}$$

Notice that opposite operations are used. Divide both sides by -2 and reverse the sign.

$$\frac{-2x}{-2} \leq \frac{-12}{-2}$$

$$x \leq 6$$

In set builder notation, $\{x: x \leq 6\}$.

Graphing on a Number Line

Integers and real numbers can be represented on a **number line.** The point on this line associated with each number is called the **graph** of the number. Notice that number lines are spaced equally, or proportionately (Figure 5).

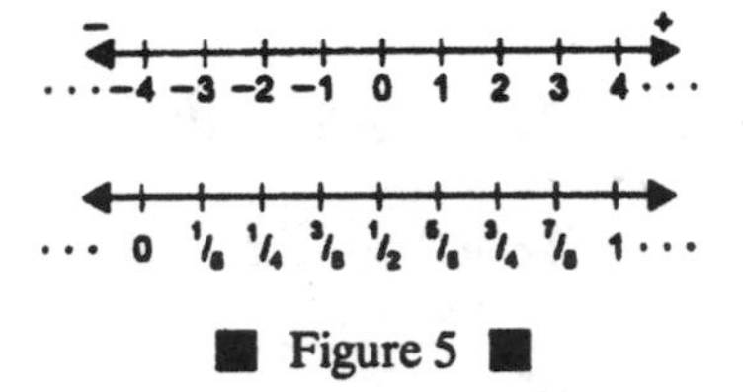

■ Figure 5 ■

Graphing inequalities. When *graphing inequalities involving only integers,* **dots** are used.

Example 4: Graph the set of x such that $1 \leq x \leq 4$ and x is an integer.

$$\{x: 1 \leq x \leq 4, x \text{ is an integer}\}$$

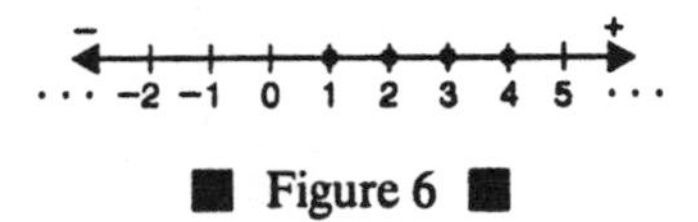

■ Figure 6 ■

When *graphing inequalities involving real numbers,* lines, rays, and dots are used. A dot is used if the number is included. A **hollow dot** is used if the number is not included.

Example 5: Graph as indicated.

(a) Graph the set of x such that $x \geq 1$.

$\{x: x \geq 1\}$

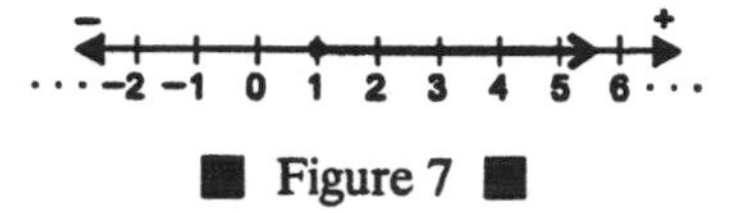

■ Figure 7 ■

(b) Graph the set of x such that $x > 1$.

$\{x: x > 1\}$

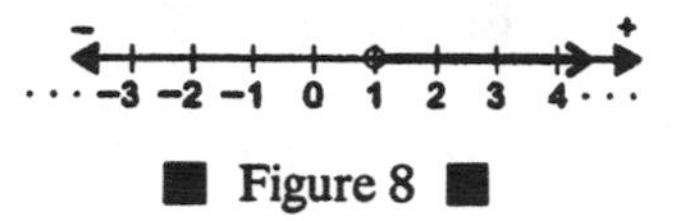

■ Figure 8 ■

(c) Graph the set of x such that $x < 4$.

$\{x: x < 4\}$

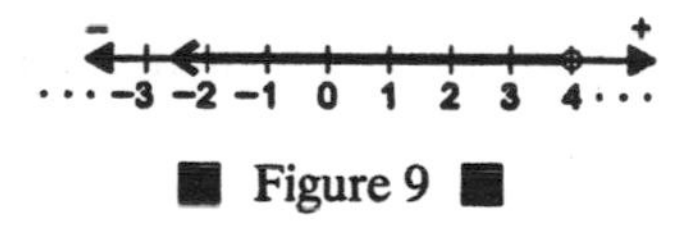

■ Figure 9 ■

This ray is often called an **open ray** or a **half line**. The hollow dot distinguishes an open ray from a ray.

Intervals. An **interval** consists of all the numbers that lie within two certain boundaries. If the two boundaries, or fixed numbers, are included, then the interval is called a **closed interval.** If the fixed numbers are not included, then the interval is called an **open interval.**

Example 6: Graph

(a) Closed interval $\{x: -1 \leq x \leq 2\}$

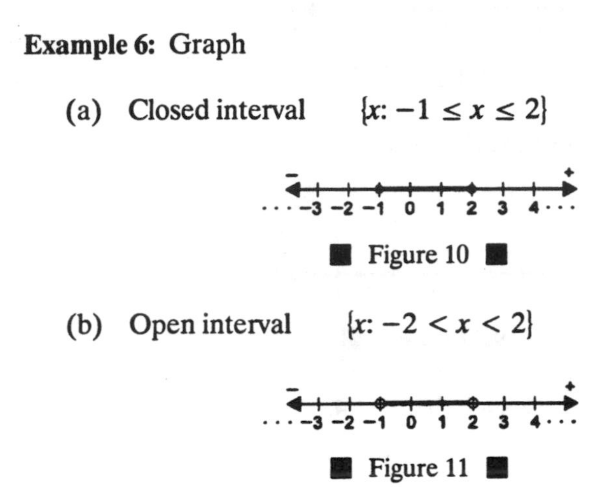

■ Figure 10 ■

(b) Open interval $\{x: -2 < x < 2\}$

■ Figure 11 ■

If the interval includes only one of the boundaries, then it is called a **half-open interval.**

Example 7: Graph the half-open interval.

$$\{x: -1 < x \leq 2\}$$

...−3 −2 −1 0 1 2 3 4...

■ Figure 12 ■

Absolute Value

The numerical value when direction or sign is not considered is called the **absolute value.** The absolute value of x is written $|x|$. The absolute value of a number is always positive except when the number is 0.

$$|0| = 0 \qquad |x| > 0 \qquad |-x| > 0$$

Example 8: Give the value.

(a) $|4| = 4$

(b) $|-6| = 6$

(c) $|7 - 9| = |-2| = 2$

(d) $3 - |-6| = 3 - 6 = -3$

Note that absolute value is taken first.

Solving equations containing absolute value. To *solve an equation containing absolute value,* isolate the absolute value on one side of the equation. Then set its contents equal to both + and − the other side of the equation and solve both equations.

Example 9: Solve $|x| + 2 = 5$.

Isolate the absolute value

$$\begin{array}{rl} |x| + 2 = & 5 \\ -2 = & -2 \\ \hline |x| \quad = & 3 \end{array}$$

Set the contents of the absolute value portion equal to +3 and −3.

$$x = 3 \qquad x = -3$$

Answer: 3, −3

Example 10: Solve $3|x - 1| - 1 = 11$.

Isolate the absolute value.

$$\begin{array}{rl} 3|x - 1| - 1 = & 11 \\ +1 \quad & +1 \\ \hline 3|x - 1| \quad = & 12 \end{array}$$

$$\frac{3|x - 1|}{3} = \frac{12}{3}$$

$$|x - 1| = 4$$

Set the contents of the absolute value portion equal to +4 and −4.

Solving for x,

$$\begin{array}{rl} x - 1 = & 4 \\ +1 \quad & +1 \\ \hline x \quad = & 5 \end{array} \qquad \begin{array}{rl} x - 1 = & -4 \\ +1 = & +1 \\ \hline x \quad = & -3 \end{array}$$

Answer: 5, −3

Solving inequalities containing absolute value and graphing. To *solve an inequality containing absolute value,* follow the same steps as in solving equations with absolute value, except you must remember to reverse the direction of the sign when setting the absolute value opposite the negative.

Example 11: Solve and graph the answer: $|x - 1| > 2$.

Isolate the absolute value.

$$|x - 1| > 2$$

Set the contents of the absolute value portion to both 2 and −2. Be sure to change the direction of the sign when using −2.

Solve for x.

$$\begin{array}{rcl} x - 1 > 2 & & x - 1 < -2 \\ +1 \quad +1 & & +1 \quad +1 \\ \hline x \quad > 3 & \text{or} & x \quad < -1 \end{array}$$

Graph the answer (Figure 13).

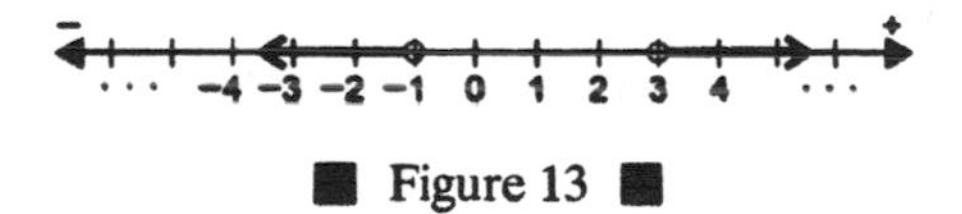

■ Figure 13 ■

Example 12: Solve and graph the answer: $3|x| - 2 \leq 1$.

Isolate the absolute value.

$$\begin{array}{rcr} 3|x| - 2 & \leq & 1 \\ +2 & & +2 \\ \hline 3|x| & \leq & 3 \end{array}$$

$$\frac{3|x|}{3} \leq \frac{3}{3}$$

$$|x| \leq 1$$

Set the contents of the absolute value portion to both 1 and -1. Be sure to change the direction of the sign when using -1.

$$x \leq 1 \quad \text{and} \quad x \geq -1$$

Graph the answer (Figure 14).

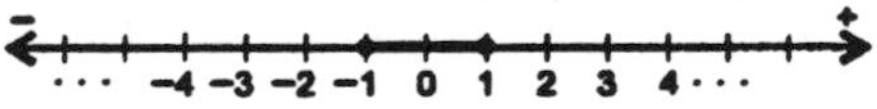

$\cdots$ −4 −3 −2 −1 0 1 2 3 4 $\cdots$

■ Figure 14 ■

Example 13: Solve and graph the answer: $2|1 - x| + 1 \geq 3$.

Isolate the absolute value.

$$\begin{array}{rcr} 2|1 - x| + 1 & \geq & 3 \\ -1 & & -1 \\ \hline 2|1 - x| & \geq & 2 \end{array}$$

$$\frac{2|1 - x|}{2} \geq \frac{2}{2}$$

$$|1 - x| \geq 1$$

Set the contents of the absolute value portion to both 1 and -1. Be sure to change the direction of the sign when using -1.

Solve for x.

$$\begin{array}{rr} 1 - x \geq 1 \\ -1 \quad\quad -1 \\ \hline -x \geq 0 \end{array} \qquad \begin{array}{rr} 1 - x \leq -1 \\ -1 \quad\quad -1 \\ \hline -x \leq -2 \end{array}$$

$$\frac{-x}{-1} \geq \frac{0}{-1} \qquad \frac{-x}{-1} \leq \frac{-2}{-1}$$

$$x \leq 0 \quad \text{or} \quad x \geq 2$$

Graph the answer (Figure 15).

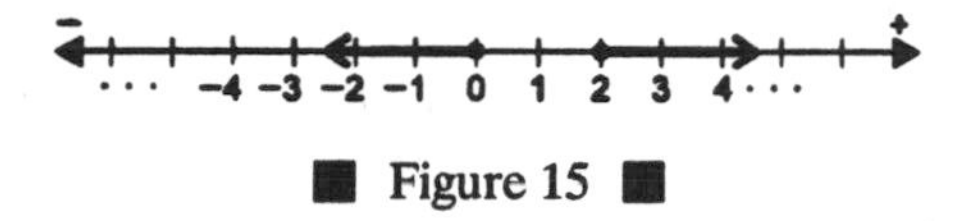

■ Figure 15 ■

Coordinate Graphs

Each point on a number line is assigned a number. In the same way, each point in a plane is assigned a pair of numbers. These numbers represent the placement of the point relative to two intersecting lines. In **coordinate graphs** (Figure 16), two perpendicular number lines are used and are called **coordinate axes.** One axis is horizontal and is called the ***x*-axis.** The other is vertical and is called the ***y*-axis.** The point of intersection of the two number lines is called the **origin** and is represented by the coordinates (0, 0).

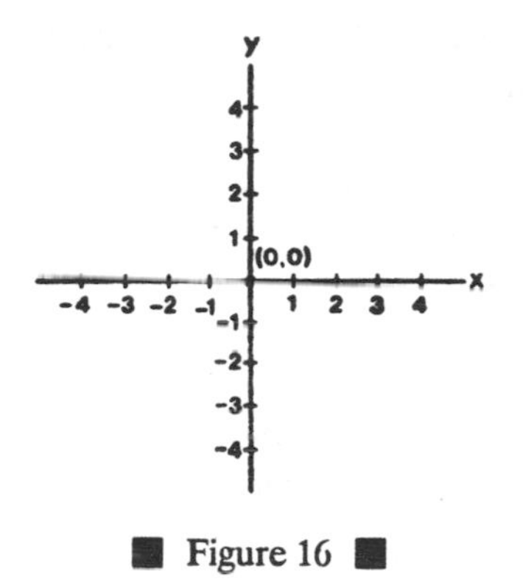

■ Figure 16 ■

Each point on a plane is located by a unique ordered pair of numbers called the **coordinates.** Some coordinates are noted in Figure 17.

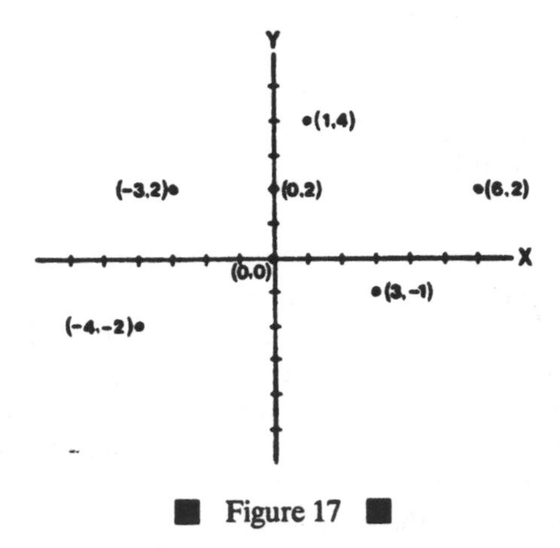

■ Figure 17 ■

Notice that on the x-axis numbers to the right of 0 are positive and to the left of 0 are negative. On the y-axis, numbers above 0 are positive and below 0 are negative. Also, note that the first number in the ordered pair is called the ***x*-coordinate,** or **abscissa,** while the second number is the ***y*-coordinate,** or **ordinate.** The x-coordinate shows the right or left direction, and the y-coordinate shows the up or down direction.

The coordinate graph is divided into four quarters called **quadrants.** These quadrants are labeled in Figure 18.

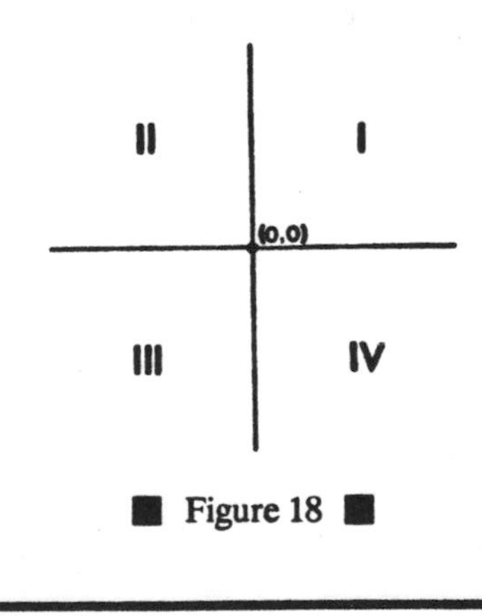

■ Figure 18 ■

Notice that

In quadrant I, x is always positive and y is always positive.
In quadrant II, x is always negative and y is always positive.
In quadrant III, x and y are both always negative.
In quadrant IV, x is always positive and y is always negative.

Graphing equations on the coordinate plane. To *graph an equation on the coordinate plane,* find the solutions by giving a value to one variable and solving the resulting equation for the other value. Repeat this process to find other solutions. (When giving a value for one variable, start with 0; then try 1, etc.) Then graph the solutions.

Example 1: Graph the equation $x + y = 6$.

If x is 0, then y is 6.

$$(0) + y = 6$$
$$y = 6$$

If x is 1, then y is 5.

$$\begin{array}{r} (1) + y = 6 \\ -1 \qquad -1 \\ \hline y = 5 \end{array}$$

If x is 2, then y is 4.

$$\begin{array}{r} (2) + y = 6 \\ -2 \qquad -2 \\ \hline y = 4 \end{array}$$

Using a simple chart is helpful.

x	y
0	6
1	5
2	4

Now, plot these coordinates as shown in Figure 19.

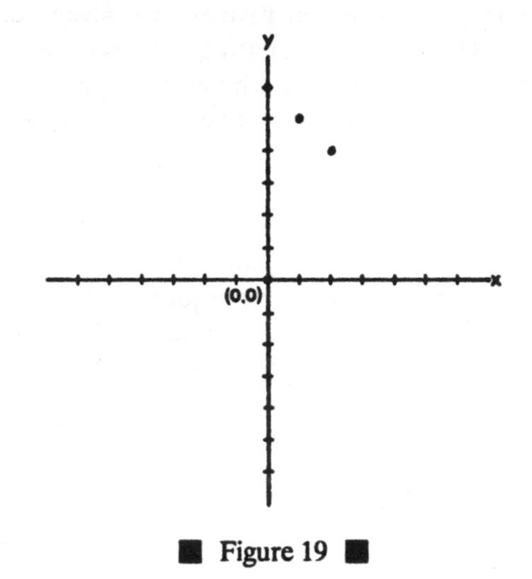

■ Figure 19 ■

Notice that these solutions, when plotted, form a straight line. Equations whose graphs of their solution sets form a straight line are called **linear equations.** Equations that have a variable raised to a power, show division by a variable, involve variables with square roots, or have variables multiplied together will not form a straight line when their solutions are graphed. These are called **nonlinear equations.**

Example 2: Graph the equation $y = x^2 + 4$.

If x is 0, then y is 4. $y = (0)^2 + 4$

$y = 0 + 4$

$y = 4$

If x is 1, then y is 5.

$$y = (1)^2 + 4$$
$$y = 1 + 4$$
$$y = 5$$

If x is 2, then y is 8.

$$y = (2)^2 + 4$$
$$y = 4 + 4$$
$$y = 8$$

Use a simple chart.

x	y
0	4
1	5
2	8

Now, plot these coordinates as shown in Figure 20.

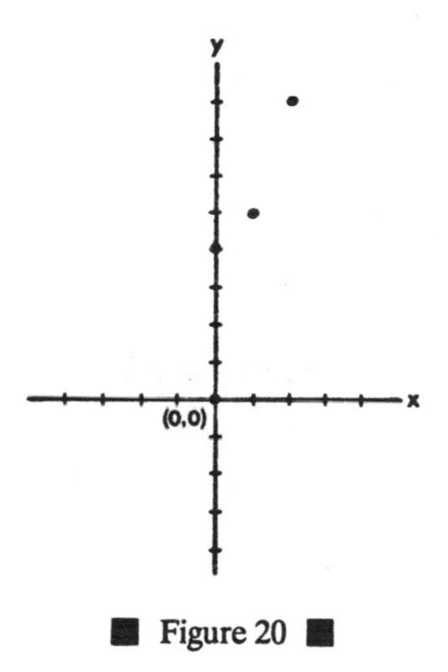

■ Figure 20 ■

Notice that these solutions, when plotted, give a curved line (nonlinear). The more points plotted, the easier it is to see and describe the solution set.

Slope and intercept of linear equations. There are two relationships between the graph of a linear equation and the equation itself that must be pointed out. One involves the **slope of the line,** and the other involves the point where the **line crosses the *y*-axis.** In order to see either of these relationships, the terms of the equation must be in a certain order.

$$(+)(1)y = (\ )x + (\ )$$

When the terms are written in this order, the equation is said to be in y-form. *Y*-form is written $\boldsymbol{y = mx + b}$, and the two relationships involve m and b.

Example 3: Write the equations in y-form.

(a) $x - y = 3$

$$-y = -x + 3$$

$$y = x - 3$$

(b) $y = -2x + 1$ (already in y form)

(c) $x - 2y = 4$

$$-2y = -x + 4$$

$$2y = x - 4$$

$$y = \frac{1}{2}x - 2$$

As shown in the graphs of these three problems in Figure 21, the lines cross the y-axis at -3, $+1$, and -2, the last term in each equation.

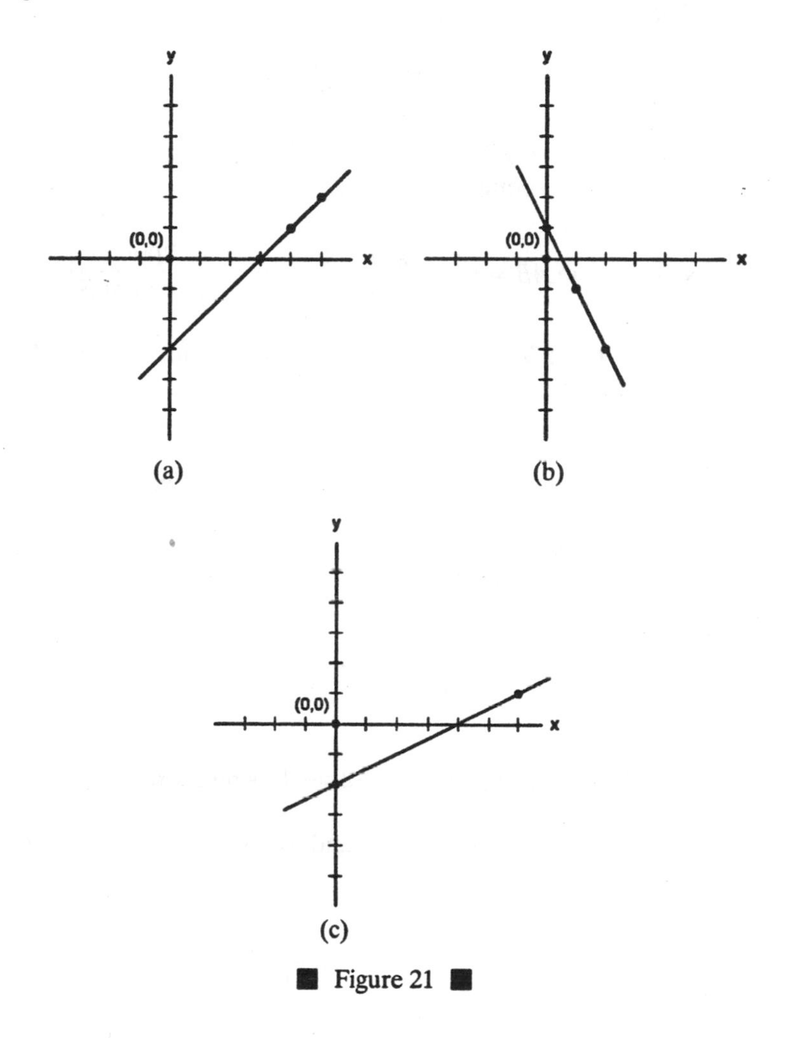

■ Figure 21 ■

If a linear equation is written in the form of $y = mx + b$, b is the y-intercept.

The **slope** of a line is defined as

$$\frac{\text{the change in } y}{\text{the change in } x}$$

and the word "change" refers to the difference in the value of y(or x) between two points on the line.

$$\text{slope of line } AB = \frac{y_A - y_B}{x_A - x_B}\left[\frac{y \text{ at point } A - y \text{ at point } B}{x \text{ at point } A - x \text{ at point } B}\right]$$

Note: Points A and B can be any two points on a line; there will be no difference in the slope.

Example 4: Find the slope of $x - y = 3$ using coordinates.

To find the slope of the line, pick any two points on the line, such as $A(3, 0)$ and $B(5, 2)$, and calculate the slope

$$\text{slope} = \frac{y_A - y_B}{x_A - x_B} = \frac{(0) - (2)}{(3) - (5)} = \frac{-2}{-2} = 1$$

Example 5: Find the slope of $y = -2x - 1$ using coordinates.

Pick two points, such as $A(1, -3)$ and $B(-1, 1)$, and calculate the slope

$$\text{slope} = \frac{y_A - y_B}{x_A - x_B} = \frac{(-3) - (1)}{(1) - (-1)} = \frac{-3 - 1}{1 + 1} = \frac{-4}{2} = -2$$

Example 6: Find the slope of $x - 2y = 4$ using coordinates.

Pick two points, such as $A(0, -2)$ and $B(4, 0)$, and calculate the slope.

$$\text{slope} = \frac{y_A - y_B}{x_A - x_B} = \frac{(-2) - (0)}{(0) - (4)} = \frac{-2}{-4} = \frac{1}{2}$$

Looking back at the equations for Example 3(a), (b), and (c) written in y-form, it should be evident that the slope of the line is the same as the numerical coefficient of the x term.

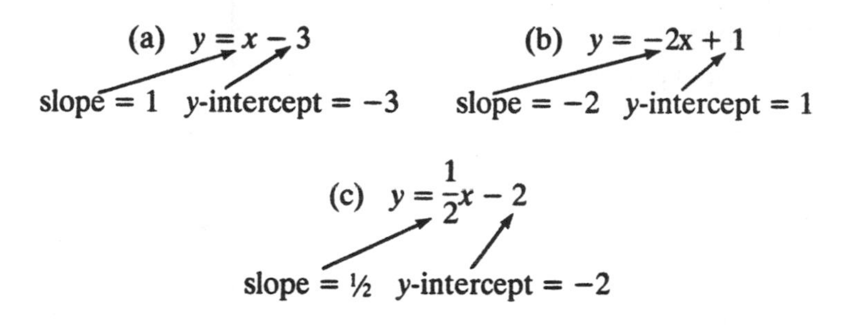

Graphing linear equations using slope and intercept. Graphing an equation by using its slope and y-intercept is usually quite easy.

1. State the equation in y-form.
2. Locate the y-intercept on the graph (that is, one of the points on the line).
3. Write the slope as a ratio (fraction) and use it to locate other points on the line.
4. Draw the line through the points.

Example 7: Graph the equation $x - y = 2$ using slope and y-intercept.

$$x - y = 2$$
$$-y = -x + 2$$
$$y = x - 2$$

Locate -2 on the y-axis and from this point, count as shown in Figure 22:

slope = 1

or $\frac{1}{1}$ (for every 1 up) (go 1 to the right)

or $\frac{-1}{-1}$ (for every 1 down) (go 1 to the left)

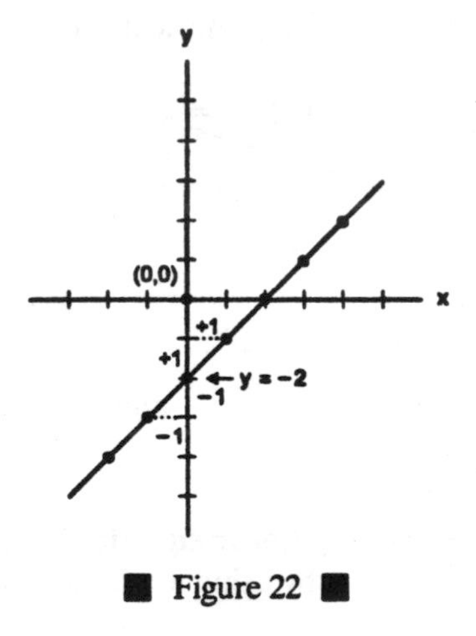

Figure 22

Example 8: Graph the equation $2x - y = -4$ using slope and y-intercept.

$$2x - y = -4$$
$$-y = -2x - 4$$
$$y = 2x + 4$$

Locate +4 on the y-axis and from this point, count as shown in Figure 23:

slope = 2

or $\frac{2}{1}$ (for every 2 up) (go 1 to the right)

or $\frac{-2}{-1}$ (for every 2 down) (go 1 to the left)

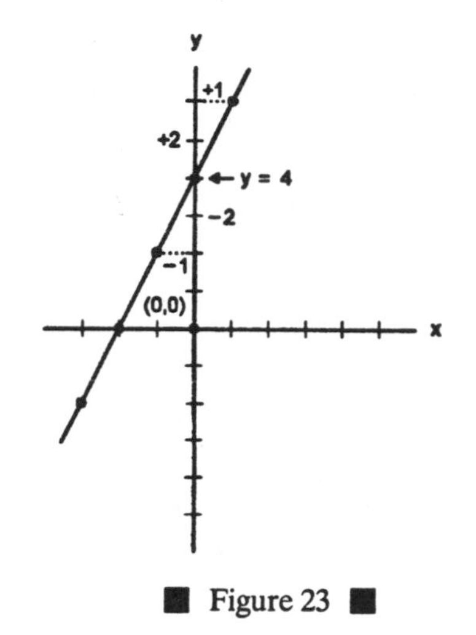

■ Figure 23 ■

Example 9: Graph the equation $x + 3y = 0$ using slope and y-intercept.

$$x + 3y = 0$$

$$3y = -x + (0)$$

$$y = -\frac{1}{3}x + (0)$$

Locate 0 on the y-axis and from this point, count as shown in Figure 24:

slope $= -\frac{1}{3}$

or $\frac{-1}{3}$ (for every 1 down) (go 3 to the right)

or $\frac{1}{-3}$ (for every 1 up) (go 3 to the left)

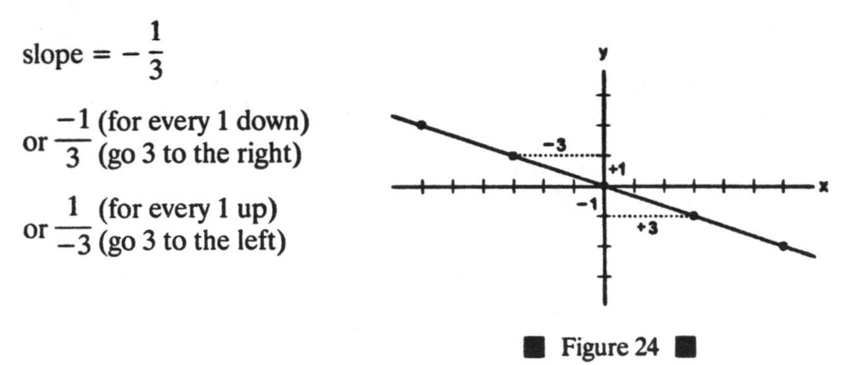

Figure 24

Finding the equation of a line. To find the equation of a line when working with ordered pairs, slopes, and intercepts, use the following approach.

1. Find the slope, m.
2. Find the y-intercept, b.
3. Substitute the slope and intercept into the slope-intercept form, $y = mx + b$.
4. Change the slope-intercept form to standard form, $Ax + By = C$.

Example 10: Find the equation of the line when $m = -4$ and $b = 3$.

1. Find the slope, m. $\quad m = -4$ (given)

2. Find the y-intercept, b. $\quad b = 3$ (given)

3. Substitute the slope and intercept into the slope-intercept form $y = mx + b$.

$$y = -4x + 3$$

4. Change the slope-intercept form to standard form $Ax + By = C$.

Since $\qquad y = -4x + 3$

adding $4x$ to each side gives

$$4x + y = 3$$

Example 11: Find the equation of the line passing through the point (6, 4) with a slope of -3.

1. Find the slope, m. $\qquad m = -3$ (given)

2. Find the y-intercept, b.

Substitute $m = -3$ and the point (6, 4) into the slope-intercept form to find b.

$$y = mx + b \text{ where } y = 4, m = -3, x = 6$$

$$4 = (-3)(6) + b$$

$$4 = -18 + b$$

$$18 + 4 = b$$

$$22 = b$$

3. Substitute the slope and intercept into the slope-intercept form $y = mx + b$.

Since $\qquad m = -3$

and $\qquad b = 22$

then $\qquad y = -3x + 22$

4. **Change the slope-intercept form to standard form $Ax + By = C$.**

   Since $y = -3x + 22$

   adding $3x$ to each side gives

   $$3x + y = 22$$

Example 12: Find the equation of the line passing through points (5, −4) and (3, −2).

1. Find the slope, m. $m = \frac{\text{change in } y}{\text{change in } x}$

   $$m = \frac{(-4) - (-2)}{5 - 3} = \frac{-4 + 2}{2} = \frac{-2}{2}$$

   $$m = -1$$

2. Find the y-intercept, b.

   Substitute the slope and either point into slope-intercept form.

   $$y = mx + b \text{ where } m = -1, x = 5, y = -4$$

   $$-4 = (-1)(5) + b$$

   $$-4 = -5 + b$$

   $$5 + -4 = b$$

   $$1 = b$$

3. Substitute the slope and intercept into the slope-intercept form $y = mx + b$.

   Since $\quad m = -1$

   and $\quad b = 1$

   then $\quad y = -1x + 1$ or $y = -x + 1$

4. Change the slope-intercept form to standard form $Ax + By = C$.

   Since $\quad y = -x + 1$

   adding x to each side gives

   $$x + y = 1$$

Linear Inequalities and Half-Planes

Each line plotted on a coordinate graph divides the graph (or plane) into two **half-planes.** This line is called the **boundary line** (or **bounding line**). The graph of a linear inequality is always a half-plane. Before graphing a linear inequality, you must first find or use the equation of the line to make a boundary line.

Open half-plane. If the inequality is a "$>$" or "$<$", then the graph will be an **open half-plane.** An open half-plane does *not* include the boundary line, so the boundary line is written as a *dashed line* on the graph.

Example 13: Graph the inequality $y < x - 3$.

First graph the line $y = x - 3$ to find the boundary line (use a dashed line, since the inequality is "$<$") as shown in Figure 25.

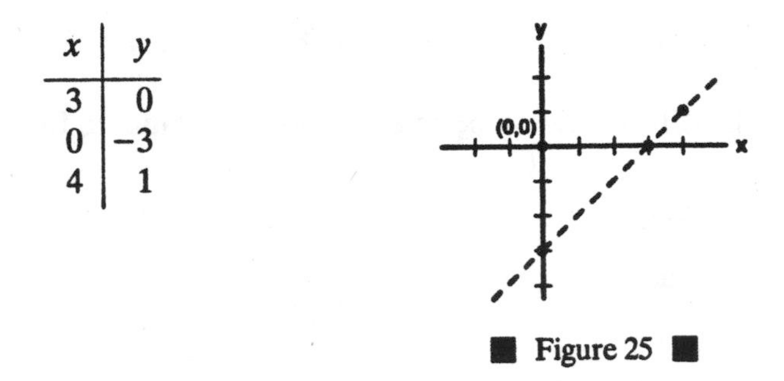

x	y
3	0
0	−3
4	1

■ Figure 25 ■

Now, *shade the lower half-plane* as shown in Figure 26, since $y < x - 3$.

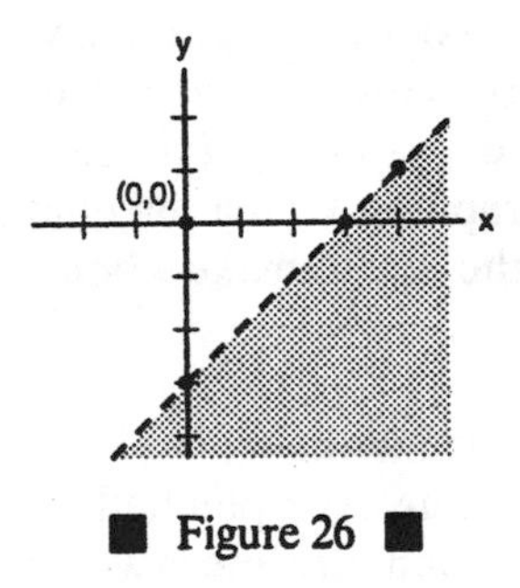

■ Figure 26 ■

To check to see if you've shaded the correct half-plane, plug in a pair of coordinates—the pair of (0, 0) is often a good choice. If the coordinates you selected make the *inequality a true statement* when plugged in, then you *should* shade the half-plane *containing* those coordinates. If the coordinates you selected *do not* make the inequality a true statement, then shade the half-plane *not containing* those coordinates.

Since the point (0, 0) *does not* make this inequality a true statement,

$y < x - 3$
$0 < 0 - 3$ is not true

you should shade the side that *does not contain* the point (0, 0).

This *checking method is often simply used as the method to decide which half-plane to shade.*

Closed half-plane. If the inequality is a "≤" or "≥", then the graph will be a **closed half-plane.** A closed half-plane includes the boundary line and is graphed using a *solid line and shading.*

Example 14: Graph the inequality $2x - y \leq 0$.

First transform the inequality so that y is the left member.

Subtracting $2x$ from each side gives

$$-y \leq -2x$$

Now dividing each side by -1 (and changing the direction of the inequality) gives

$$y \geq 2x$$

Graph $y = 2x$ to find the boundary (use a solid line, since the inequality is "$\geq$") as shown in Figure 27.

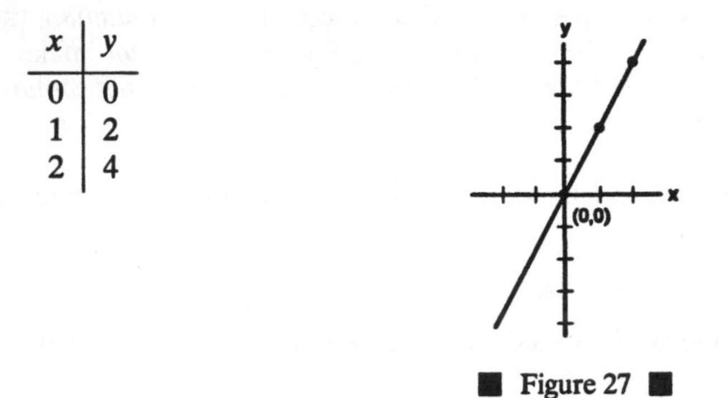

x	y
0	0
1	2
2	4

■ Figure 27 ■

Since $y \geq 2x$, you should shade the upper half-plane. If in doubt, or to check, plug in a pair of coordinates. Try the pair (1, 1).

$$y \geq 2x$$

$$1 \geq 2(1)$$

$$1 \geq 2 \quad \text{is not true}$$

So you should shade the half-plane that *does not contain* (1, 1) as shown in Figure 28.

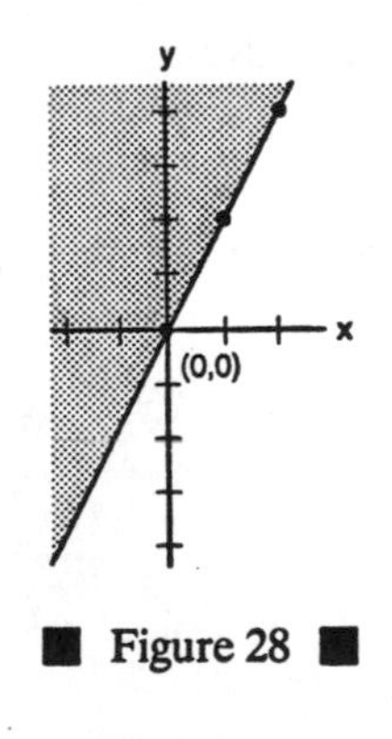

■ Figure 28 ■

Functions

Relations. Any set of ordered pairs is called a **relation.** Figure 29 shows a set of ordered pairs.

$$A = \{(-1, 1)(1, 3)(2, 2)(3,4)\}$$

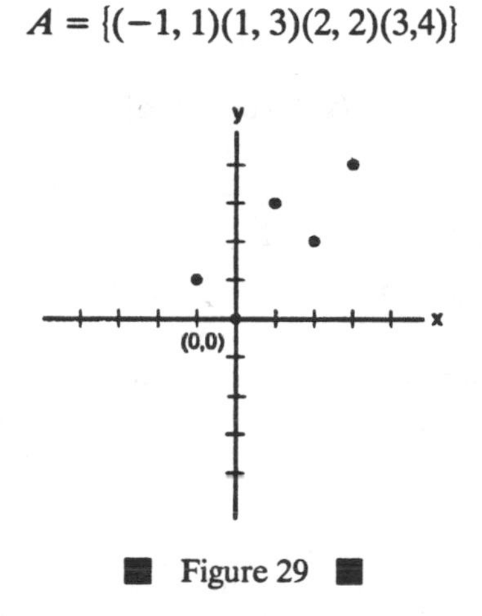

Figure 29

Domain and range. The set of all x's is called the **domain** of the relation. The set of all y's is called the **range** of the relation. The domain of set A in Figure 29 is $\{-1, 1, 2, 3\}$, while the range of set A is $\{1, 3, 2, 4\}$.

Example 1: Find the domain and range of the set of graphed points in Figure 30.

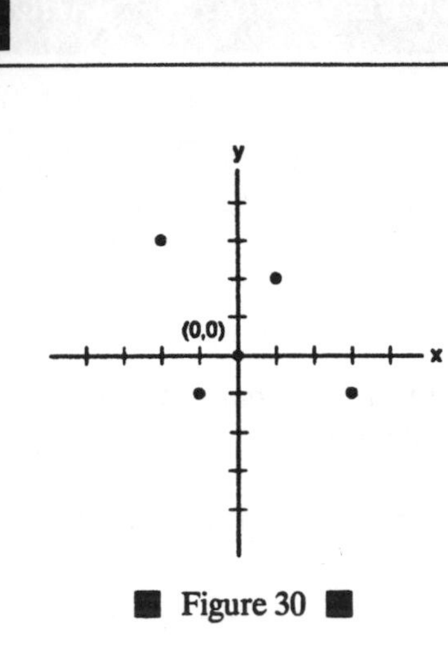

■ Figure 30 ■

The domain is the set $\{-2, -1, 1, 3\}$. The range is the set $\{3, -1, 2, -1\}$ or simply $\{3, -1, 2\}$.

Defining a function. The relation in Example 1 has pairs of coordinates with unique first terms. When the x value of each pair of coordinates is different, the relation is called a **function.** A function is a relation in which each member of the domain is paired with exactly one element of the range. *All functions are relations, but not all relations are functions.* A good example of a functional relation can be seen in the linear equation $y = x + 1$, graphed in Figure 31. The domain and range of this function are both the set of real numbers, and the relation is a function because for any value of x there is a unique value of y.

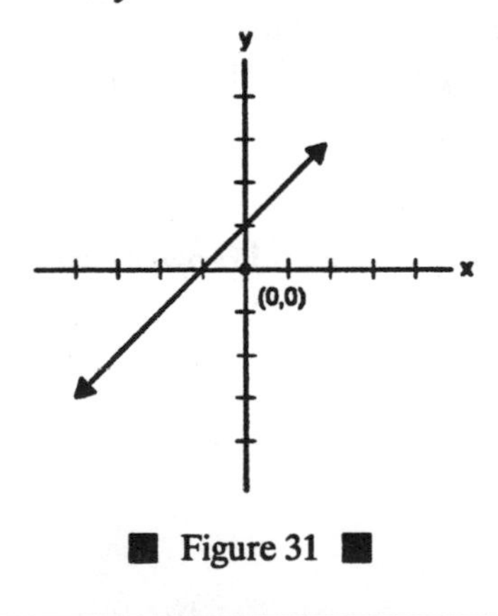

■ Figure 31 ■

Graphs of functions. In each case in Figure 32(a), (b), and (c), for any value of x, there is only one value for y. Contrast this with the graphs in Figure 33.

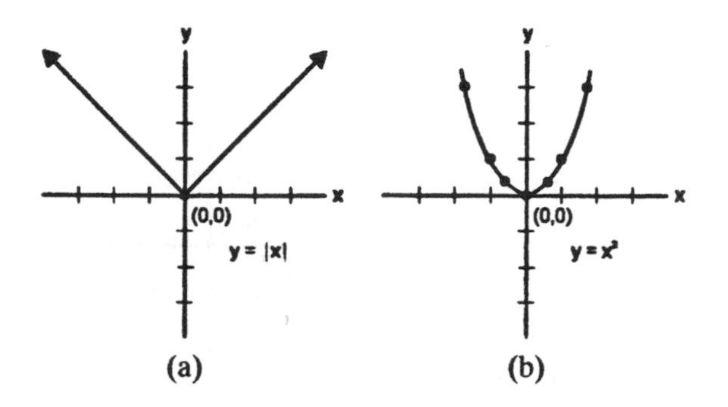

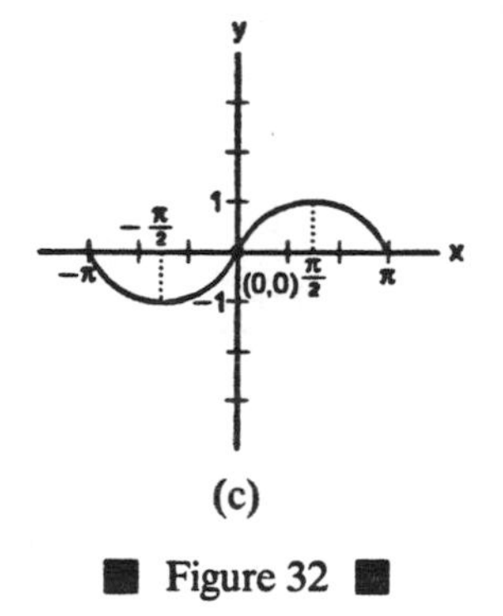

■ Figure 32 ■

Graphs of relationships that are not functions. In each of the relations in Figure 33(a), (b), and (c), a single value of x is associated with two or more values of y. These relations are *not* functions.

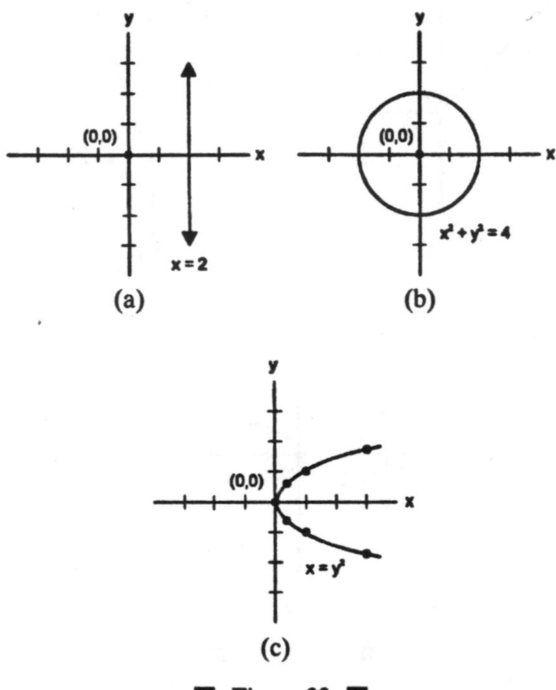

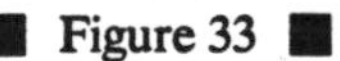
Figure 33

Determining domain, range, and if the relation is a function.

Example 2:

(a) $B = \{(-2, 3)(-1, 4)(0, 5)(1, -3)\}$ domain: $\{-2, -1, 0, 1\}$
range: $\{3, 4, 5, -3\}$
function: yes

(b)

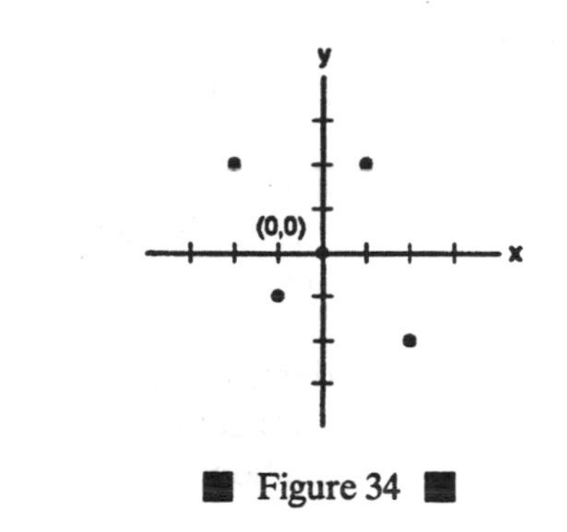

■ Figure 34 ■

domain: $\{-2, -1, 1, 2\}$
range: $\{-2, -1, 2\}$
function: yes

(c)

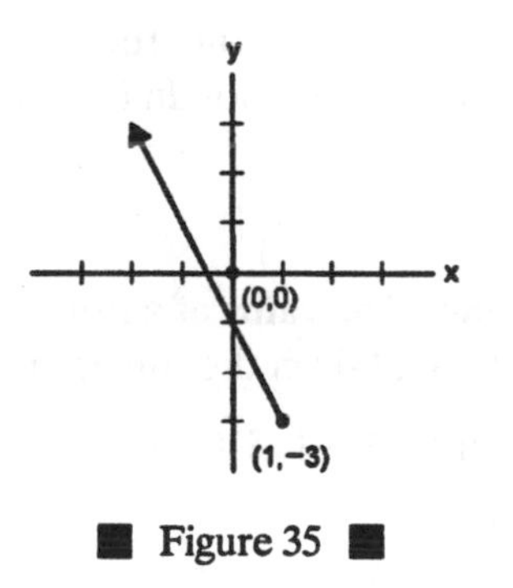

■ Figure 35 ■

domain: $\{x: x \leq 1\}$
range: $\{y: y \geq -3\}$
function: yes

(d)

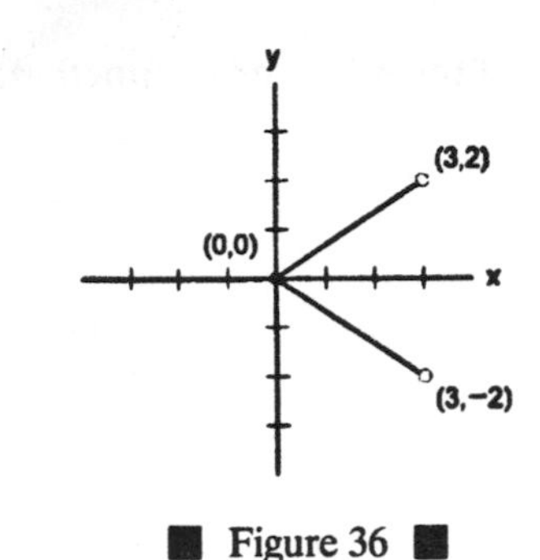

domain: $\{x: 0 \leq x < 3\}$
range: $\{y: -2 < y < 2\}$
function: no

■ Figure 36 ■

(e) $y = x^2$

domain: {all real numbers}
range: $\{y: y \geq 0\}$
function: yes

(f) $x = y^2$

domain: $\{x: x \geq 0\}$
range: {all real numbers}
function: no

Note that Example 2(e) and (f) are illustrations of **inverse relations:** relations where the domain and the range have been interchanged. Notice that while the relation in (e) is a function, the inverse relation in (f) is not.

Finding the values of functions. The **value of a function** is really the **value of the range** of the relation. Given the function

$$f = \{(1, -3)(2, 4)(-1, 5)(3, -2)\}$$

the value of the function at 1 is −3, at 2 is 4, etc. This is written $f(1) = -3$ and $f(2) = 4$ and is usually read, "f of $1 = -3$ and f of $2 = 4$." The lower-case letter f has been used here to indicate the concept of function, but any lower-case letter might have been used.

Example 3: Let $h = \{(3, 1)(2, 2)(1, -2)(-2, 3)\}$. Find each of the following.

(a) $h(3) =$ (b) $h(2) =$ (c) $h(1) =$ (d) $h(-2) =$

$h(3) = 1$ $h(2) = 2$ $h(1) = -2$ $h(-2) = 3$

Example 4: If $g(x) = 2x + 1$, find each of the following.

(a) $g(-1) =$

$$g(x) = 2x + 1$$
$$g(-1) = 2(-1) + 1$$
$$g(-1) = -2 + 1$$
$$g(-1) = -1$$

(b) $g(2) =$

$$g(x) = 2x + 1$$
$$g(2) = 2(2) + 1$$
$$g(2) = 4 + 1$$
$$g(2) = 5$$

(c) $g(a) =$

$$g(x) = 2x + 1$$
$$g(a) = 2(a) +$$
$$g(a) = 2a + 1$$

Example 5: If $f(x) = 3x^2 + x - 1$, find the range of f for the domain $\{1, -2, -1\}$.

$$f(x) = 3x^2 + x - 1$$
$$f(1) = 3(1)^2 + (1) - 1$$
$$f(1) = 3(1) + 1 - 1$$
$$f(1) = 3$$

$$f(x) = 3x^2 + x - 1$$
$$f(-2) = 3(-2)^2 + (-2) - 1$$
$$f(-2) = 3(4) - 2 - 1$$
$$f(-2) = 12 - 3$$
$$f(-2) = 9$$

$$f(x) = 3x^2 + x - 1$$
$$f(-1) = 3(-1)^2 + (-1) - 1$$
$$f(-1) = 3(1) - 1 - 1$$
$$f(-1) = 3 - 2$$
$$f(-1) = 1$$

range: {3, 9, 1}

Variations

A variation is a relation between a set of values of one variable and a set of values of other variables.

Direct variation. In the equation $y = mx + b$, if m is a nonzero constant and $b = 0$, then you have the function $y = mx$ (often written $y = kx$), which is called a **direct variation.** That is, you can say that *y varies directly as x or y is directly proportional to x.* In this function, m (or k) is called the **constant of proportionality** or the **constant of variation.** *The graph of every direct variation passes through the origin.*

Example 6: Graph $y = 2x$.

x	y
0	0
1	2
2	4

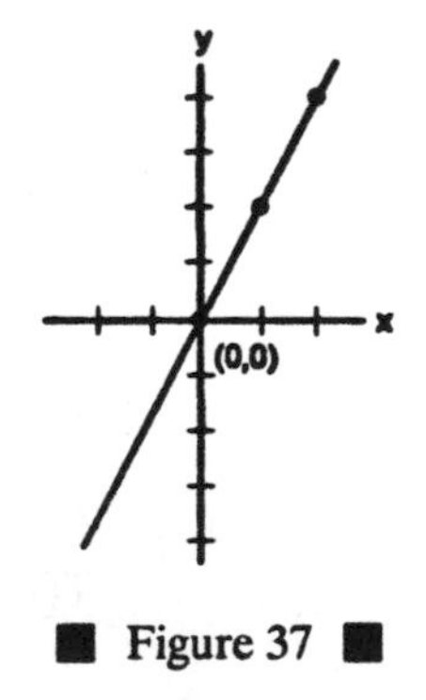

■ Figure 37 ■

Example 7: If y varies directly as x, find the constant of variation when y is 2 and x is 4.

Since this is a direct variation,

$$y = kx \quad (\text{or } y = mx)$$

Now, replacing y with 2 and x with 4,

$$2 = k(4)$$

So $$k = 2/4 \text{ or } 1/2$$

The constant of variation is $1/2$.

Example 8: If y varies directly as x and the constant of variation is 2, find y when x is 6.

Since this is a direct variation, simply replace k with 2 and x with 6 in the following equation.

$$y = kx$$

$$y = 2(6)$$

So $$y = 12$$

A direct variation can also be written as a proportion.

$$\frac{y_1}{x_1} = \frac{y_2}{x_2}$$

This proportion is read, "y_1 is to x_1 as y_2 is to x_2." x_1 and y_2 are called the **means,** and y_1 and x_2 are called the **extremes.** The product of the means is always equal to the product of the extremes. You can solve a proportion by simply multiplying the means and extremes and then solving as usual.

Example 9: *r* varies directly as *p*. If *r* is 3 when *p* is 7, find *p* when *r* is 9.

Set up the direct proportion

$$\frac{r_1}{p_1} = \frac{r_2}{p_2}$$

Now, substitute in the values.

$$\frac{3}{7} = \frac{9}{p}$$

Multiply the means and extremes (cross multiplying) gives

$$3p = 63$$

So

$$p = 21$$

Inverse variation (indirect variation). A variation where $y = m/x$ or $y = k/x$ is called an **inverse variation** (or **indirect variation**). That, is, *as x increases, y decreases.* And *as y increases, x decreases.* You may see the equation $xy = k$ representing an inverse variation, but this is simply a rearrangement of $y = k/x$. This function is also referred to as an **inverse** or **indirect proportion.** Again, m (or k) is called the constant of variation.

Example 10: If y varies indirectly as x, find the constant of variation when y is 2 and x is 4.

Since this is an indirect or inverse variation,

$$y = k/x$$

Now, replacing y with 2 and x with 4,

$$2 = k/4$$

So $$k = 2(4) \text{ or } 8$$

The constant of variation is 8.

Example 11: If y varies indirectly as x and the constant of variation is 2, find y when x is 6.

Since this is an indirect variation, simply replace k with 2 and x with 6 in the following equation.

$$y = k/x$$

$$y = 2/6$$

So $$y = 1/3$$

Note: This subject is introduced in the Pre-Algebra section (page 12).

The symbol $\sqrt{\ }$ is called a **radical sign** and is used to designate **square root.** To designate **cube root,** a small three is placed above the radical sign, $\sqrt[3]{\ }$. When two radical signs are next to each other, they automatically mean that the two are multiplied. The multiplication sign may be omitted. Note that the square root of a negative number is not possible within the real number system. A completely different system of **imaginary numbers** is used. The (so-called) imaginary numbers are multiples of the **imaginary unit** i.

$$\sqrt{-1} = i, \quad \sqrt{-4} = 2i, \quad \sqrt{-9} = 3i, \text{ etc.}$$

Simplifying Square Roots

Example 1: Simplify.

(a) $\sqrt{9} = 3$

(b) $-\sqrt{9} = -3$

Reminder: This notation is used in many texts and will be adhered to in this book.

(c) $\sqrt{18} = \sqrt{9 \cdot 2} = \sqrt{9} \cdot \sqrt{2} = 3\sqrt{2}$

(d) If each variable is nonnegative (not a negative number),

$\sqrt{x^2} = x$

If each variable could be positive or negative (deleting the restriction "If each variable is nonnegative"), then absolute value signs are placed around variables to odd powers.

$$\sqrt{x^2} = |x|$$

(e) If each variable is nonnegative,

$\sqrt{x^4} = x^2$

(f) If each variable is nonnegative,

$\sqrt{x^6y^8} = \sqrt{x^6}\sqrt{y^8} = x^3y^4$

If each variable could be positive or negative, then you would write

$$|x^3|y^4$$

(g) If each variable is nonegative,

$\sqrt{25a^4b^6} = \sqrt{25}\sqrt{a^4}\sqrt{b^6} = 5a^2b^3$

If each variable could be positive or negative, you would write

$5a^2|b^3|$

(h) If each variable is nonegative,

$\sqrt{x^7} = \sqrt{x^6(x)} = \sqrt{x^6}\sqrt{x} = x^3\sqrt{x}$

If each variable could be positive or negative, you would write

$$|x^3|\sqrt{x}$$

(i) If each variable is nonnegative,

$$\sqrt{x^9y^8} = \sqrt{x^9}\sqrt{y^8} = \sqrt{x^8(x)}\sqrt{y^8} = x^4\sqrt{x} \cdot y^4 = x^4y^4\sqrt{x}$$

(j) If each variable is nonnegative,

$$\sqrt{16x^5} = \sqrt{16}\sqrt{x^5} = \sqrt{16}\sqrt{x^4(x)} = 4x^2\sqrt{x}$$

Operations with Square Roots

Under a single radical sign. You may perform *operations under a single radical sign.*

Example 2: Perform the operation indicated

(a) $\sqrt{(5)(20)} = \sqrt{100} = 10$

(b) $\sqrt{30 + 6} = \sqrt{36} = 6$

(c) $\sqrt{\frac{32}{2}} = \sqrt{16} = 4 \left(\text{Note: } \sqrt{\frac{32}{2}} = \frac{\sqrt{32}}{\sqrt{2}}\right)$

(d) $\sqrt{30 - 5} = \sqrt{25} = 5$

(e) $\sqrt{2 + 5} = \sqrt{7}$

When radical values are alike. You can *add or subtract square roots themselves only if the values under the radical sign are equal.* Then simply add or subtract the coefficients (numbers in front of the radical sign) and keep the original number in the radical sign.

Example 3: Perform the operation indicated.

(a) $2\sqrt{3} + 3\sqrt{3} = (2 + 3)\sqrt{3} = 5\sqrt{3}$

(b) $4\sqrt{6} - 2\sqrt{6} = (4 - 2)\sqrt{6} = 2\sqrt{6}$

(c) $5\sqrt{2} + \sqrt{2} = 5\sqrt{2} + 1\sqrt{2} = (5 + 1)\sqrt{2} = 6\sqrt{2}$

Note that 1 is understood in $\sqrt{2}$. ($1\sqrt{2}$)

When radical values are different. You *may not add or subtract different square roots.*

Example 4:

(a) $\sqrt{28} - \sqrt{3} \neq \sqrt{25}$

(b) $\sqrt{16} + \sqrt{9} \neq \sqrt{25}$

Addition and subtraction of square roots after simplifying. Sometimes, after *simplifying the square root(s), addition or subtraction becomes possible.* Always simplify if possible.

Example 5: Simplify and add.

(a) $\sqrt{50} + 3\sqrt{2} =$

These cannot be added until $\sqrt{50}$ is simplified.

$$\sqrt{50} = \sqrt{25 \cdot 2} = \sqrt{25} \cdot \sqrt{2} = 5\sqrt{2}$$

Now, since both are alike under the radical sign,

$$5\sqrt{2} + 3\sqrt{2} = (5 + 3)\sqrt{2} = 8\sqrt{2}$$

(b) $\sqrt{300} + \sqrt{12} =$

Try to simplify each one.

$$\sqrt{300} = \sqrt{100 \cdot 3} = \sqrt{100} \cdot \sqrt{3} = 10\sqrt{3}$$

$$\sqrt{12} = \sqrt{4 \cdot 3} = \sqrt{4} \cdot \sqrt{3} = 2\sqrt{3}$$

Now, since both are alike under the radical sign,

$$10\sqrt{3} + 2\sqrt{3} = (10 + 2)\sqrt{3} = 12\sqrt{3}$$

Products of nonnegative roots. Remember that in multiplication of roots, the multiplication sign may be omitted. Always simplify the answer when possible.

Example 6: Multiply.

(a) $\sqrt{2} \cdot \sqrt{8} = \sqrt{16} = 4$

(b) If each variable is nonnegative,

$\sqrt{x^3} \cdot \sqrt{x^5} = \sqrt{x^8} = x^4$

(c) If each variable is nonnegative,

$$\sqrt{ab} \cdot \sqrt{ab^3c} = \sqrt{a^2b^4c} = \sqrt{a^2}\sqrt{b^4}\sqrt{c} = ab^2\sqrt{c}$$

(d) If each variable is nonnegative,

$$\sqrt{3x} \cdot \sqrt{6xy^2} \cdot \sqrt{2xy} = \sqrt{36x^3y^3} = \sqrt{36}\sqrt{x^3}\sqrt{y^3} =$$

$$\sqrt{36}\sqrt{x^2(x)}\sqrt{y^2(y)} = 6xy\sqrt{xy}$$

(e) $2\sqrt{5} \cdot 7\sqrt{3} = (2 \cdot 7)\sqrt{5 \cdot 3} = 14\sqrt{15}$

Quotients of nonnegative roots. For all positive numbers,

$$\frac{\sqrt{x}}{\sqrt{y}} = \sqrt{\frac{x}{y}}$$

In the following examples, all variables are assumed to be positive.

Example 7: Divide. Leave all fractions with rational denominators.

(a) $\frac{\sqrt{10}}{\sqrt{2}} = \sqrt{\frac{10}{2}} = \sqrt{5}$

(b) $\frac{\sqrt{24}}{\sqrt{3}} = \sqrt{\frac{24}{3}} = \sqrt{8} = 2\sqrt{2}$

(c) $\frac{\sqrt{28x^6}}{\sqrt{7x^2}} = \sqrt{\frac{28x^6}{7x^2}} = \sqrt{4x^4} = 2x^2$

(d) $\frac{\sqrt{15}}{\sqrt{6}} = \sqrt{\frac{15}{6}} = \sqrt{\frac{5}{2}}$ or $\frac{\sqrt{5}}{\sqrt{2}}$

Note that the denominator of this fraction is irrational. In order to rationalize the denominator of this fraction, multiply it by 1 in the form of

$$\frac{\sqrt{2}}{\sqrt{2}}$$

$$\frac{\sqrt{5}}{\sqrt{2}}\cdot 1=\frac{\sqrt{5}}{\sqrt{2}}\cdot\frac{\sqrt{2}}{\sqrt{2}}=\frac{\sqrt{10}}{\sqrt{4}}=\frac{\sqrt{10}}{2}$$

Example 8: Divide. Leave all fractions with rational denominators.

(a) $\frac{5\sqrt{7}}{\sqrt{12}}$

First simplify $\sqrt{12}$:

$$\frac{5\sqrt{7}}{\sqrt{12}}=\frac{5\sqrt{7}}{2\sqrt{3}}\cdot 1=\frac{5\sqrt{7}}{2\sqrt{3}}\cdot\frac{\sqrt{3}}{\sqrt{3}}=\frac{5\sqrt{21}}{2\cdot 3}=\frac{5\sqrt{21}}{6}$$

or

$$\frac{5\sqrt{7}}{\sqrt{12}}\cdot\frac{\sqrt{12}}{\sqrt{12}}=\frac{5\sqrt{7}\cdot\sqrt{12}}{12}=\frac{5\sqrt{84}}{12}=\frac{5\sqrt{4\cdot 21}}{12}=\frac{10\sqrt{21}}{12}=\frac{5\sqrt{21}}{6}$$

(b) $\frac{9\sqrt{2x}}{\sqrt{24x^3}}=9\sqrt{\frac{2x}{24x^3}}=\frac{9}{\sqrt{12x^2}}=\frac{9}{2x\sqrt{3}}\cdot 1=\frac{9}{2x\sqrt{3}}\cdot\frac{\sqrt{3}}{\sqrt{3}}=$

$$\frac{9\sqrt{3}}{2x\cdot 3}=\frac{9\sqrt{3}}{6x}=\frac{3\sqrt{3}}{2x}$$

(c) $\frac{3}{2+\sqrt{3}} \cdot 1 = \frac{3}{(2+\sqrt{3})} \cdot \frac{(2-\sqrt{3})}{(2-\sqrt{3})} = \frac{3(2-\sqrt{3})}{4-3} = \frac{6-3\sqrt{3}}{1} =$

$6 - 3\sqrt{3}$

Note: In order to leave an irrational term in the denominator, it is necessary to multiply both the numerator and denominator by the **conjugate** of the denominator. The conjugate of a binomial contains the same terms but the opposite sign. Thus, $(x + y)$ and $(x - y)$ are conjugates.

Example 9: Divide. Leave the fraction with a rational denominator.

$$\frac{1+\sqrt{5}}{2-\sqrt{5}} \cdot 1 = \frac{(1+\sqrt{5})}{(2-\sqrt{5})} \cdot \frac{(2+\sqrt{5})}{(2+\sqrt{5})} = \frac{2+3\sqrt{5}+5}{4-5} =$$

$$\frac{7+3\sqrt{5}}{-1} = -7 - 3\sqrt{5}$$

Solving Quadratic Equations

A **quadratic equation** is an equation that could be written as

$$ax^2 + bx + c = 0$$

There are three basic methods for solving quadratic equations: factoring, using the quadratic formula, and completing the square.

Factoring. To solve a quadratic equation by factoring,

1. Put all terms on one side of the equal sign, leaving zero on the other side.
2. Factor.
3. Set each factor equal to zero.
4. Solve each of these equations.
5. Check by inserting your answer in the original equation.

Example 1: Solve $x^2 - 6x = 16$.

Following the steps,

$$x^2 - 6x \text{ becomes } x^2 - 6x - 16 = 0$$

Factor

$$(x - 8)(x + 2) = 0$$

$$x - 8 = 0 \quad \text{or} \quad x + 2 = 0$$

$$x = 8 \qquad\qquad x = -2$$

Then to check,

$$8^2 - 6(8) = 16 \quad \text{or} \quad (-2)^2 - 6(-2) = 16$$
$$64 - 48 = 16 \qquad\qquad 4 + 12 = 16$$
$$16 = 16 \qquad\qquad 16 = 16$$

Both values, 8 and -2, are solutions to the original equation.

Example 2: Solve $y^2 = -6y - 5$.

Setting all terms equal to zero,

$$y^2 + 6y + 5 = 0$$

Factor.

$$(y + 5)(y + 1) = 0$$

Setting each factor to 0,

$$y + 5 = 0 \quad \text{or} \quad y + 1 = 0$$
$$y = -5 \qquad\qquad y = -1$$

To check,

$$(-5)^2 = -6(-5) - 5 \quad \text{or} \quad (-1)^2 = -6(-1) - 5$$
$$25 = 30 - 5 \qquad\qquad 1 = 6 - 5$$
$$25 = 25 \qquad\qquad 1 = 1$$

A quadratic with a term missing is called an **incomplete quadratic.**

Example 3: Solve $x^2 - 16 = 0$.

Factor.

$$(x + 4)(x - 4) = 0$$

$$x + 4 = 0 \quad \text{or} \quad x - 4 = 0$$

$$x = -4 \qquad\qquad x = 4$$

To check,

$$(-4)^2 - 16 = 0 \quad \text{or} \quad (4)^2 - 16 = 0$$

$$16 - 16 = 0 \qquad\qquad 16 - 16 = 0$$

$$0 = 0 \qquad\qquad 0 = 0$$

Example 4: Solve $x^2 + 6x = 0$.

Factor.

$$x(x + 6) = 0$$

$$x = 0 \quad \text{or} \quad x + 6 = 0$$

$$x = 0 \qquad\qquad x = -6$$

To check,

$$(0)^2 + 6(0) = 0 \quad \text{or} \quad (-6)^2 + 6(-6) = 0$$

$$0 + 0 = 0 \qquad\qquad 36 + (-36) = 0$$

$$0 = 0 \qquad\qquad 0 = 0$$

Example 5: Solve $2x^2 + 2x - 1 = x^2 + 6x - 5$.

First, simplify by putting all terms on one side and combining like terms.

$$\begin{array}{r} 2x^2 + 2x - 1 = x^2 + 6x - 5 \\ -x^2 - 6x + 5 = -x^2 - 6x + 5 \\ \hline x^2 - 4x + 4 = 0 \end{array}$$

Now, factor.

$$(x - 2)(x - 2) = 0$$
$$x - 2 = 0$$
$$x = 2$$

To check,

$$2(2)^2 + 2(2) - 1 = (2)^2 + 6(2) - 5$$
$$8 + 4 - 1 = 4 + 12 - 5$$
$$11 = 11$$

The quadratic formula. Many quadratic equations cannot be solved by factoring. This is generally true when the roots, or answers, are not rational numbers. A second method of solving quadratic equations involves the use of the formula

$$x = \frac{-b \pm \sqrt{b^2 - 4ac}}{2a}$$

where a, b, and c are taken from the quadratic equation written in its general form of

$$ax^2 + bx + c = 0$$

where a is the numeral that goes in front of x^2, b is the numeral that goes in front of x, and c is the numeral with no variable next to it.

When using the quadratic formula, you should be aware of three possibilities. These three possibilities are distinguished by a part of the formula called the **discriminant.** The discriminant is the value under the radical sign $b^2 - 4ac$. A quadratic equation with real numbers as coefficients can have

1. two different real roots if the discriminant $b^2 - 4ac$ is a positive number
2. one real root if the discriminant $b^2 - 4ac$ is equal to 0
3. no real root if the discriminant $b^2 - 4ac$ is a negative number

Example 6: Solve for x: $x^2 - 5x = -6$.

Setting all terms equal to 0,

$$x^2 - 5x + 6 = 0$$

Then substitute 1 (which is understood to be in front of the x^2), -5, and 6 for a, b, and c in the quadratic formula and simplify.

$$x = \frac{-b \pm \sqrt{b^2 - 4ac}}{2a}$$

$$x = \frac{-(-5) \pm \sqrt{(-5)^2 - 4(1)(6)}}{2(1)}$$

$$x = \frac{5 \pm \sqrt{25 - 24}}{2}$$

$$x = \frac{5 \pm \sqrt{1}}{2}$$

$$x = \frac{5 \pm 1}{2}$$

$$x = \frac{5+1}{2} \qquad x = \frac{5-1}{2}$$

$$x = \frac{6}{2} \qquad x = \frac{4}{2}$$

$$x = 3 \qquad x = 2$$

Since the discriminant $b^2 - 4ac$ is positive, you get two different real roots.

Example 6 produces rational roots. Following, the quadratic formula is used to solve an equation whose roots are not rational.

Example 7: Solve for y: $y^2 = -2y + 2$.

Setting all terms equal to 0,

$$y^2 + 2y - 2 = 0$$

Then substitute 1, 2, and -2 for a, b, and c in the quadratic formula and simplify.

$$y = \frac{-b \pm \sqrt{b^2 - 4ac}}{2a}$$

$$y = \frac{-(2) \pm \sqrt{(2)^2 - 4(1)(-2)}}{2(1)}$$

$$y = \frac{-2 \pm \sqrt{4+8}}{2}$$

$$y = \frac{-2 \pm \sqrt{12}}{2}$$

$$y = \frac{-2 \pm \sqrt{4}\sqrt{3}}{2}$$

$$y = \frac{-2 \pm 2\sqrt{3}}{2}$$

$$y = \frac{2(-1 \pm \sqrt{3})}{2}$$

$$y = -1 + \sqrt{3} \qquad y = -1 - \sqrt{3}$$

Note that the two roots are irrational.

Example 8: Solve for x: $x^2 + 2x + 1 = 0$.

Substituting in the quadratic formula,

$$x = \frac{-b \pm \sqrt{b^2 - 4ac}}{2a}$$

$$x = \frac{-2 \pm \sqrt{(2)^2 - (4)(1)(1)}}{2(1)}$$

$$x = \frac{-2 \pm \sqrt{4 - 4}}{2}$$

$$x = \frac{-2 \pm \sqrt{0}}{2}$$

$$x = \frac{-2}{2} = -1$$

Since the discriminant $b^2 - 4ac$ is 0, the equation has one root.

The quadratic formula can also be used to solve quadratic equations whose roots are imaginary numbers, that is, they have no solution in the real number system.

Example 9: Solve for x: $x(x + 2) + 2 = 0$, or $x^2 + 2x + 2 = 0$.

Substituting in the quadratic formula,

$$x = \frac{-b \pm \sqrt{b^2 - 4ac}}{2a}$$

$$x = \frac{-(2) \pm \sqrt{(2)^2 - 4(1)(2)}}{2(1)}$$

$$x = \frac{-2 \pm \sqrt{4 - 8}}{2}$$

$$x = \frac{-2 \pm \sqrt{-4}}{2}$$

Since the discriminant $b^2 - 4ac$ is negative, this equation has no solution in the real number system.

Completing the square. A third method of solving quadratic equations that works with both real and imaginary roots is called completing the square.

1. Put the equation in the form of $ax^2 + bx = -c$.
2. Make sure that $a = 1$ (if $a \neq 1$, multiply through the equation by $1/a$ before proceeding).
3. Using the value of b from this new equation, add $(b/2)^2$ to both sides of the equation to form a perfect square on the left side of the equation.
4. Find the square root of both sides of the equation.
5. Solve the resulting equation.

Example 10: Solve for x: $x^2 - 6x + 5 = 0$.

Arrange in the form of

$$ax^2 + bx = -c$$

$$x^2 - 6x = -5$$

Since $a = 1$, add $(-6/2)^2$, or 9, to both sides to complete the square.

$$x^2 - 6x + 9 = -5 + 9$$

$$x^2 - 6x + 9 = 4$$

or $$(x - 3)^2 = 4$$

Take the square root of both sides.

$$x - 3 = \pm 2$$

Solve.

$$x = 3 \pm 2$$

$$x = 3 + 2 \quad x = 3 - 2$$

$$x = 5 \qquad x = 1$$

Example 11: Solve for y: $y^2 + 2y - 4 = 0$.

Arrange in the form of

$$ay^2 + by = -c$$

$$y^2 + 2y = 4$$

Since $a = 1$, add $(2/2)^2$, or 1, to both sides to complete the square.

$$y^2 + 2y + 1 = 4 + 1$$

$$y^2 + 2y + 1 = 5$$

or $$(y + 1)^2 = 5$$

Take the square root of both sides.

$$y + 1 = \pm\sqrt{5}$$

Solve.

$$y = -1 \pm \sqrt{5}$$

$$y = -1 + \sqrt{5} \qquad y = -1 - \sqrt{5}$$

Example 12: Solve for x: $2x^2 + 3x + 2 = 0$.

Arrange in the form of

$$ax^2 + bx = -c$$

$$2x^2 + 3x = -2$$

Since $a \neq 1$, multiply through the equation by ½.

$$x^2 + \frac{3}{2}x = -1$$

Add $[(1/2)(3/2)]^2$, or $9/16$, to both sides.

$$x^2 + \frac{3}{2}x + \frac{9}{16} = -1 + \frac{9}{16}$$

$$x^2 + \frac{3}{2}x + \frac{9}{16} = -\frac{7}{16}$$

$$\left(x + \frac{3}{4}\right)^2 = -\frac{7}{16}$$

Take the square root of both sides.

$$x + \frac{3}{4} = \pm \sqrt{\frac{-7}{16}}$$

$$x + \frac{3}{4} = \pm \frac{\sqrt{-7}}{\sqrt{16}}$$

$$x + \frac{3}{4} = \frac{\pm\sqrt{-7}}{4}$$

$$x = \frac{-3}{4} \pm \frac{\sqrt{-7}}{4}$$

There is no solution in the real number system.

Word problems are often the nemesis of even the best math student. For many, the difficulty is not the computation. The problems stem from what is given and what is being asked.

Solving Technique

There are many types of word problems involving arithmetic, algebra, geometry, and combinations of each with various twists. It is most important to have a systematic technique for solving word problems. Here is such a technique.

1. First, *identify what is being asked.* What are you ultimately trying to find? How far a car has traveled? How fast a plane flies? How many items can be purchased? Whatever it is, find it and then *circle it.* This helps insure that you are solving for what is being asked.
2. Next, *underline and pull out information you are given* in the problem. Draw a picture if you can. This helps you know what you have and will point you to a relationship or equation. Note any key words in the problem (see "Key Words and Phrases" following).
3. If you can, *set up an equation or some straightforward system* with the given information.
4. *Is all the given information necessary to solve the problem?* Occasionally, you may be given more than enough information to solve a problem. *Choose what you need* and don't spend needless energy on irrelevant information.

5. *Carefully solve the equation or work the necessary computation.* Be sure you are working in the same units (for example, you may have to change feet into inches, pounds into ounces, etc., in order to keep everything consistent).
6. *Did you answer the question?* One of the most common errors in answering word problems is the failure to answer what was actually being asked.
7. And finally, *is your answer reasonable?* Check to make sure that an error in computation or a mistake in setting up your equation did not give you a ridiculous answer.

Key Words and Phrases

In working with word problems, there are some words or phrases that give clues as to how the problem should be solved. The most common words or phrases are as follows.

- **Add**

  Sum—as in *the sum of 2, 3, and 6 . . .*
  Total—as in *the total of the first six payments . . .*
  Addition—as in *a recipe calls for the addition of five pints . . .*
  Plus—as in *three liter plus two liters . . .*
  Increase—as in *her pay was increased by $15 . . .*
  More than—as in *this week the enrollment was eight more than last week . . .*
  Added to—as in *if you added $3 to the cost . . .*

- **Subtract**

  Difference—as in *what is the difference between . . .*
  Fewer—as in *there were fifteen fewer men than women . . .*
  Remainder—as in *how many are left or what quantity remains . . .*
  Less than—as in *a number is five less than another number . . .*
  Reduced—as in *the budget was reduced by $5,000 . . .*
  Decreased—as in *if he decreased the speed of his car by ten miles per hour . . .*
  Minus—as in *some number minus 9 is . . .*

- **Multiply**

  Product—as in *the product of 8 and 5 is . . .*
  Of—as in *one-half of the group . . .*
  Times—as in *five times as many girls as boys . . .*
  At—as in *the cost of ten yards of material at 70¢ a yard is . . .*
  Total—as in *if you spend $15 a week on gas, what is the total for a three-week period . . .*
  Twice—as in *twice the value of some number . . .*

- **Divide**

  Quotient—as in *the final quotient is . . .*
  Divided by—as in *some number divided by 12 is . . .*
  Divided into—as in *the group was divided into . . .*
  Ratio—as in *what is the ratio of . . .*
  Half—as in *half the profits are . . .* (dividing by 2)

  As you work a variety of word problem types, you will discover more "clue" words.

A final remainder: Be sensitive to what each of these questions is asking. What time? How many? How much? How far? How old? What length? What is the ratio?

Simple Interest

Example 1: How much simple interest will an account earn in five years if \$500 is invested at 8% interest per year?

First, circle what you must find—*interest.* Now use the equation

Interest = *principal* times *rate* times *time*

$$I = prt$$

Simply plug into the equation.

$$I = \$500(.08)5$$
$$I = \$200$$

Note that both rate and time are in yearly terms (annual rate; years).

Compound Interest

Example 2: What will be the final total after three years on an original investment of \$1,000 if a 12% annual interest rate is compounded yearly?

First, circle what you must find—*final total.* Note also that interest will be *compounded each year.* Therefore, the solution has three parts, one for each year.

total for first year: *Interest* = *principal* times *rate* times *time*

$$I = prt$$
$$I = \$1{,}000 \times .12 \times 1$$
$$I = \$120$$

Thus, the *total* after one year is 1,000 + \$120 = \$1,120.

total for second year: $I = prt$

$I = \$1{,}120(.12)1$

$I = \$134.40$

Note that the principal at the beginning of the second year was \$1,120. Thus, the total after two years is \$1,120 + \$134.40 = \$1,254.40.

total for third year: $I = prt$

$I = \$1{,}254.40(.12)1$

$I = \$150.53$

Note that the principal at the beginning of the second year was \$1,254.40. Thus, the total after three years is \$1,254.40 + \$150.53 = \$1,404.93.

Ratio and Proportion

Example 3: If Arnold can type 600 pages of manuscript in twenty-one days, how many days will it take him to type 230 pages if he works at the same rate?

First, circle what you're asked to find—*how many days.* One simple way to work this problem is to set up a "framework" (proportion) using the categories given in the equation. Here the categories are *pages* and *days.* Therefore, a framework may be

$$\frac{\text{pages}}{\text{days}} = \frac{\text{pages}}{\text{days}}$$

Note that you also may have used

$$\frac{\text{days}}{\text{pages}} = \frac{\text{days}}{\text{pages}}$$

The answer will still be the same. Now, simply plug into the equation for each instance.

$$\frac{600}{21} = \frac{230}{x}$$

Cross multiplying,

$$600x = 21(230)$$

$$600x = 4{,}830$$

$$\frac{600x}{600} = \frac{4{,}830}{600}$$

$$x = 8\tfrac{1}{20} \quad \text{or} \quad 8.05$$

Therefore, it will take $8\frac{1}{20}$ or 8.05 days to type 230 pages. (You may have simplified the original proportion before solving.)

Percent

Example 4: Thirty students are awarded doctoral degrees at the graduate school, and this number comprises 40% of the total graduate student body. How many graduate students were enrolled?

First, circle what you must find in the problem—*how many graduate students.* Now, in order to plug into the percentage equation

$$\frac{\text{is}}{\text{of}} = \%$$

try rephrasing the question into a simple sentence. For example, in this case,

30 is 40% of what total?

Notice that the 30 sits next to the word *is*; therefore, 30 is the "is" number. 40 is the percent. Notice that *what total* sits next to the word *of*. Therefore, plugging into the equation,

$$\frac{\text{is}}{\text{of}} = \%$$

$$\frac{30}{x} = \frac{40}{100}$$

Cross multiplying,

$$40x = 3{,}000$$

$$\frac{40x}{40} = \frac{3{,}000}{40}$$

$$x = 75$$

Thereforc, the total graduate enrollment was 75 students.

Percent Change

To find the *percent change* (increase or decrease), use the formula given in Example 5.

Example 5: Last year, Harold earned \$250 a month at his after-school job. This year, his after-school earnings have increased to \$300 per month. What is the percent increase in his monthly after-school earnings?

First, circle what you're looking for—*percent increase.* Percent change (percent increase, percentage rise, % difference, percent decrease, etc.) is always found by using the equation

$$\text{percent change} = \frac{\text{change}}{\text{starting point}}$$

Therefore,

$$\begin{aligned}\text{percent change} &= \frac{\$300 - \$250}{\$250}\\ &= \frac{\$50}{\$250}\\ &= \frac{1}{5} = .20 = 20\%\end{aligned}$$

The percent increase in Harold's after-school salary is 20%.

Number Problems

Example 6: When 6 times a number is increased by 4, the result is 40. Find the number.

First, circle what you must find—*the number.* Letting x stand for the number gives the equation

$$6x + 4 = 40$$

Subtracting 4 from each side gives

$$6x = 36$$

Dividing by 6 gives

$$x = 6$$

So the number is 6.

Example 7: One number exceeds another number by 5. If the sum of the two numbers is 39, find the smaller number.

First, circle what you are looking for—*the smaller number.* Now, let the smaller number equal x. Therefore, the larger number equals $x + 5$. Now, use the problem to set up an equation.

$$\underbrace{\text{If the sum of the two numbers}}_{x + (x + 5)} \underbrace{\text{is}}_{=} \underbrace{39}_{39} \ldots$$

$$2x + 5 = 39$$

$$2x + 5 - 5 = 39 - 5$$

$$2x = 34$$

$$\frac{2x}{2} = \frac{34}{2}$$

$$x = 17$$

Therefore, the smaller number is 17.

Example 8: If one number is three times as large as another number and the smaller number is increased by 19, the result is 6 less than twice the larger number. What is the larger number?

First, circle what you must find—*the larger number.* Let the smaller number equal x. Therefore, the larger number will be $3x$. Now, using the problem, set up an equation.

The smaller number increased by 19 is 6 less than twice the larger number.

$$\underbrace{x}_{\text{The smaller number}} \quad \underbrace{+}_{\text{increased by}} \quad \underbrace{19}_{19} \quad \underbrace{=}_{\text{is}} \quad \underbrace{2(3x) - 6}_{\text{6 less than twice the larger number}}$$

$$x + 19 = 6x - 6$$

$$-x + x + 19 = -x + 6x - 6$$

$$19 = 5x - 6$$

$$19 + 6 = 5x - 6 + 6$$

$$25 = 5x$$

$$5 = x$$

Therefore, the larger number, $3x$, is $3(5)$, or 15.

Example 9: The sum of three consecutive integers is 306. What is the largest integer?

First, circle what you must find—*the largest integer.* Let the smallest integer equal x; let $x + 1$ equal the next integer; let the largest integer equal $x + 2$. Now, use the problem to set up an equation.

$\underbrace{\text{The sum of three consecutive integers}}\ \underbrace{\text{is}}\ \underbrace{306}.$

$$x + (x + 1) + (x + 2) = 306$$

$$3x + 3 = 306$$

$$3x + 3 - 3 = 306 - 3$$

$$3x = 303$$

$$\frac{3x}{3} = \frac{303}{3}$$

$$x = 101$$

Therefore, the largest integer, $x + 2$, is $101 + 2$, or 103.

Age Problems

Example 10: Tom and Phil are brothers. Phil is thirty-five years old. Three years ago, Phil was four times as old as his brother was then. How old is Tom now?

First, circle what it is you must ultimately find—*Tom now.* Therefore, let t be Tom's age now. Then three years ago, Tom's age would be $t - 3$. Four times Tom's age three years ago would be $4(t - 3)$. Phil's age three years ago would be $35 - 3 = 32$. A simple chart may also be helpful.

	now	3 years ago
Phil	35	32
Tom	t	$t - 3$

Now, use the problem to set up an equation.

Three years ago Phil was four times as old as his brother was then.

$$\underbrace{\text{Phil}}_{32}\ \underbrace{\text{was}}_{=}\ \underbrace{\text{four times}}_{4\text{ times}}\ \ldots\ \underbrace{\text{brother was then}}_{(t-3)}$$

$$\frac{32}{4} = \frac{4(t-3)}{4}$$

$$8 = t - 3$$

$$8 + 3 = t - 3 + 3$$

$$11 = t$$

Therefore, Tom is now 11.

Example 11: Lisa is 16 years younger than Kathy. If the sum of their ages is 30, how old is Lisa?

First, circle what you must find—*Lisa.* Let Lisa equal x. Therefore, Kathy is $x + 16$. (Note that since Lisa is 16 years *younger* than Kathy, you must *add* 16 years to Lisa to denote Kathy's age.) Now, use the problem to set up an equation.

$$\underbrace{\text{If the sum of their ages}}_{\text{Lisa + Kathy}}\ \underbrace{\text{is}}_{=}\ \underbrace{30}_{30}\ldots$$

$$x + (x + 16) = 30$$

$$2x + 16 = 30$$

$$2x + 16 - 16 = 30 - 16$$

$$2x = 14$$

$$\frac{2x}{2} = \frac{14}{2}$$

$$x = 7$$

Therefore, Lisa is 7 years old.

Motion Problems

Example 12: How long will it take a bus traveling 72 km/hr to go 36 kms?

First circle what you're trying to find—*how long will it take* (time). Motion problems are solved by using the equation

$$distance = rate \text{ times } time$$

$$d = rt$$

Therefore, simply plug in: 72 km/hr is the rate (or speed) of the bus, and 36 km is the distance.

$$d = rt$$

$$36 \text{ km} = (72 \text{ km/hr})(t)$$

$$\frac{36}{72} = \frac{72t}{72}$$

$$\frac{1}{2} = t$$

Therefore, it will take one-half hour for the bus to travel 36 km at 72 km/hr.

Example 13: How fast in miles per hour must a car travel to go 600 miles in 15 hours?

First, circle what you must find—*how fast* (rate). Now, using the equation $d = rt$, simply plug in 600 for distance and 15 for time.

$$d = rt$$

$$600 = r(15)$$

$$\frac{600}{15} = \frac{r(15)}{15}$$

$$40 = r$$

So the rate is 40 miles per hour.

Example 14: Mrs. Benevides leaves Burbank at 9 A.M. and drives west on the Ventura Freeway at an average speed of 50 miles per hour. Ms. Twill leaves Burbank at 9:30 A.M. and drives west on the Ventura Freeway at an average speed of 60 miles per hour. At what time will Ms. Twill overtake Mrs. Benevides, and how many miles will they each have gone?

First, circle what you are trying to find—*at what time* and *how many miles.* Now, let t stand for the time Ms. Twill drives before overtaking Mrs. Benevides. Then Mrs. Benevides drives for $t + \frac{1}{2}$ hours before being overtaken. Next, set up the following chart.

	rate r ×	*time* t =	*distance* d
Ms. Twill	60 mph	t	$60t$
Mrs. Benevides	50 mph	$t + \frac{1}{2}$	$50(t + \frac{1}{2})$

Since each travels the same distance,

$$60t = 50(t + \tfrac{1}{2})$$

$$60t = 50t + 25$$

$$10t = 25$$

$$t = 2.5$$

Ms. Twill overtakes Mrs. Benevides after 2.5 hours of driving. The exact time can be figured out by using Ms. Twill's starting time: 9:30 + 2:30 = 12 noon. Since Ms. Twill has traveled for 2.5 hours at 60 mph, she has traveled 2.5×60, which is 150 miles. So Mrs. Benevides is overtaken at 12 noon, and each has traveled 150 miles.

Coin Problems

Example 15: Tamar has four more quarters than dimes. If he has a total of \$1.70, how many quarters and dimes does he have?

First, circle what you must find—*how many quarters and dimes.* Let x stand for the number of dimes, then $x + 4$ is the number of quarters. Therefore, $.10x$ is the total value of the dimes, and $.25(x + 4)$ is the total value of the quarters. Setting up the following chart can be helpful.

	number	value	amount of money
dimes	x	.10	$.10x$
quarters	$x + 4$	.25	$.25(x + 4)$

Now, use the table and problem to set up an equation.

$$.10x + .25(x + 4) = 1.70$$

$$10x + 25(x + 4) = 170$$

$$10x + 25x + 100 = 170$$

$$35x + 100 = 170$$

$$35x = 70$$

$$x = 2$$

So there are two dimes. Since there are four more quarters, there must be six quarters.

Example 16: Sid has $4.85 in coins. If he has six more nickels than dimes and twice as many quarters as dimes, how many coins of each type does he have?

First, circle what you must find—the number of *coins of each type.* Let x stand for the number of dimes. Then $x + 6$ is the number of nickels, and $2x$ is the number of quarters. Setting up the following chart can be helpful.

	number	value	amount of money
dimes	x	.10	$.10x$
nickels	$x + 6$	.05	$.05(x + 6)$
quarters	$2x$	.25	$.25(2x)$

Now, use the table and problem to set up an equation.

$$.10x + .05(x + 6) + .25(2x) = 4.85$$

$$10x + 5(x + 6) + 25(2x) = 485$$

$$10x + 5x + 30 + 50x = 485$$

$$65x + 30 = 485$$

$$65x = 455$$

$$x = 7$$

So there are seven dimes. Therefore, there are thirteen nickels and fourteen quarters.

Mixture Problems

Example 17: Coffee worth \$1.05 per pound is mixed with coffee worth 85¢ per pound to obtain twenty pounds of a mixture worth 90¢ per pound. How many pounds of each type are used?

First, circle what you are trying to find—*how many pounds of each type*. Now, let the number of pounds of \$1.05 coffee be denoted as x. Therefore, the number of pounds of 85¢-per-pound coffee must be the remainder of the twenty pounds, or $20 - x$. Now, make a chart for the cost of each type and the total cost.

	cost per lb. ×	amount in lbs. =	total of each
\$1.05 coffee	\$1.05	x	\$1.05x
\$ .85 coffee	\$ .85	$20 - x$	\$ .85$(20 - x)$
mixture	\$ .90	20	\$ .90(20)

Now, set up the equation.

$$\underbrace{\$1.05x}_{\text{Total cost of one type}} \underbrace{+}_{\text{plus}} \underbrace{.85(20 - x)}_{\text{total cost of other type}} \underbrace{=}_{\text{equal}} \underbrace{.90(20)}_{\text{total cost of mixture.}}$$

$$1.05x + 17.00 - .85x = 18.00$$

$$17.00 + .20x = 18.00$$

$$-17.00 + 17.00 + 20x = 18.00 - 17.00$$

$$.20x = 1.00$$

$$\frac{.20x}{.20} = \frac{1.00}{.20}$$

$$x = 5$$

Therefore, five pounds of coffee worth \$1.05 per pound are used. And $20 - x$, or $20 - 5$, or fifteen pounds of 85¢-per-pound coffee are used.

Example 18: Solution A is 50% hydrochloric acid, while solution B is 75% hydrochloric acid. How many liters of each solution should be used to make 100 liters of a solution which is 60% hydrochloric acid?

First, circle what you're trying to find—*liters of solutions A and B.* Now, let x stand for the number of liters of solution A. Therefore, the number of liters of solution B must be the remainder of the 100 liters, or $100 - x$. Next, make the following chart.

	% of acid	liters	concentration of acid
solution A	50%	x	$.50x$
solution B	75%	$100 - x$	$.75(100 - x)$
new solution	60%	100	$.60(100)$

Now, set up the equation.

$$.50x + .75(100 - x) = .60(100)$$

$$\begin{array}{r} .50x + 75 - .75x = 60 \\ -75 \qquad\qquad -75 \\ \hline .50x \qquad - .75x = -15 \end{array}$$

$$-.25x = -15$$

$$\frac{-.25x}{-.25} = \frac{-15}{-.25}$$

$$x = 60$$

Therefore, using the chart, 60 liters of solution A and 40 liters of solution B are used.

Work Problems

Example 19: Ernie can plow a field alone in four hours. It takes Sid five hours to plow the same field alone. If they work together (and each has a plow), how long will it take to plow the field?

First, circle what you must find—*how long . . . together.* Work problems of this nature may be solved by using the following equation.

$$\frac{1}{\text{1st person's rate}} + \frac{1}{\text{2nd person's rate}} + \frac{1}{\text{3rd person's rate}} + \text{etc.} = \frac{1}{\text{rate together}}$$

Therefore,

$$\frac{1}{\text{Ernie's rate}} + \frac{1}{\text{Sid's rate}} = \frac{1}{\text{rate together}}$$

$$\frac{1}{4} + \frac{1}{5} = \frac{1}{t}$$

Finding a common denominator,

$$\frac{5}{20} + \frac{4}{20} = \frac{1}{t}$$

$$\frac{9}{20} = \frac{1}{t}$$

Cross multiplying,

$$9t = 20$$

$$\frac{9t}{9} = \frac{20}{9} = 2\tfrac{2}{9} \text{ hours}$$

Therefore, it will take them $2\frac{2}{9}$ hours working together.

Number Problems with Two Variables

Example 20: The sum of two numbers is 15. The difference of the same two numbers is 7. What are the two numbers?

First, circle what you're looking for—*the two numbers.* Let x stand for the larger number and y stand for the second number. Now, set up two equations.

The sum of the two numbers is 15.

$$x + y = 15$$

The difference is 7.

$$x - y = 7$$

Now, solve by adding the two equations.

$$\begin{aligned} x + y &= 15 \\ x - y &= \ \ 7 \\ \hline 2x \qquad &= 22 \end{aligned}$$

So

$$x = 11$$

Now, plugging into the first equation gives

$$11 + y = 15$$

So

$$y = \ \ 4$$

The numbers are 11 and 4.

Example 21: The sum of two numbers is 20 and their product is 96. Find the numbers.

First, circle what you must find—*the numbers.* Let x stand for one of the numbers and y stand for the other number. Now set up two equations.

The sum of the two numbers is 20.

$$x + y = 20$$

Their product is 96.

$$x(y) = 96$$

Rearranging the first equation gives

$$y = 20 - x$$

Now, substituting the first equation into the second gives

$$x(20 - x) = 96$$

$$20x - x^2 = 96$$

$$-x^2 + 20x = 96$$

Putting this equation in standard form leaves

$$x^2 - 20x + 96 = 0$$

Factoring gives

$$(x - 8)(x - 12) = 0$$

Setting each factor equal to 0 and solving gives

$$x - 8 = 0 \qquad x - 12 = 0$$

So the numbers are 8 and 12.

Using another technique, you could have set up the equation as follows. Let x equal one number; then $20 - x$ is the other number. Using the phrase "their product is 96" gives the equation $x(20 - x) = 96$.